Ivars Peterson

## FATAL DEFECT

Ivars Peterson writes on physics, computers, and mathematics for *Science News*. He is the author of the bestselling *Mathematical Tourist* as well as *Islands of Truth* and *Newton's Clock*. He lives in Washington, D.C., with his wife and two children.

*Books by* Ivars Peterson

*Newton's Clock:*
*Chaos in the Solar System*

*Islands of Truth:*
*A Mathematical Mystery Cruise*

*The Mathematical Tourist:*
*Snapshots of Modern Mathematics*

*Fatal Defect:*
*Chasing Killer Computer Bugs*

# FATAL DEFECT

## Chasing Killer Computer Bugs

*Ivars Peterson*

**VINTAGE BOOKS**

A Division of Random House, Inc.     New York

## To my wife, Nancy

FIRST VINTAGE BOOKS EDITION, APRIL 1996

*Copyright © 1995, 1996 by Ivars Peterson*

The Library of Congress has cataloged the Times Books
edition as follows:
Peterson, Ivars.
Fatal defect : chasing killer computer bugs / Ivars Peterson.
— 1st ed.
p.    cm.
Includes index.
ISBN 0-8129-2023-6
1. Computer software—Reliability.   2. Computers—
Reliability.   I. Title.
QA76.76.R44P48   1995
005.3—dc20      94-41064
Vintage ISBN: 0-679-74027-9

*Author photograph © Nancy Peterson*

Random House Web address:
http://www.randomhouse.com/

Printed in the United States of America
10  9  8  7  6  5  4  3  2

# CONTENTS

# Bug Hunt

---

**WITH A RESENTFUL SIGH,** then a determined whine, our desktop computer awakes, prodded to life by a finger on the power button. Chattering noisily, it stutters through its extensive checklist, reading what it needs from the hard drive where the necessary instructions and data are invisibly stored on a spinning disk. The monitor crackles as electronically painted characters flash into view, rapidly filling row upon row. Cryptic phrases—the alphanumeric technoshorthand of the computer age—spill down the glowing screen.

At last, all is ready, and the extravagantly hued main menu appears. Our sons, Eric and Kenneth, know what to do. One types the letter *D*, and they see a list of their favorite computer programs. On this particular Sunday morning, they choose the second item.

Rumbles, roars, and snatches of conversation abruptly emanate from stereo speakers. "I've never seen anything like it," a tinny, disembodied voice remarks, just before vivid, stylized graphics fill the screen, and the opening scenario of *Where in Space Is Carmen Sandiego?* plays itself out. With its brisk animation, sly sound effects, clever plotting, humorous touches, and detailed portraits of the planets and their satellites, this game has proved a kid pleaser. With a little help, Eric, who is nearly seven years old, has successfully completed twenty missions and earned a rank of satellite deputy. Kenneth, who is four, is content to watch this time, though from past encounters, he has more than a nodding acquaintance with the main bodies of the solar system.

As the game proceeds, tension mounts. Time is running out. We have almost identified the criminal, and we are headed for Saturn. The planet's image fills the upper part of the screen, and one by one, the clue categories blink into view. Suddenly, the picture freezes, leaving the list incomplete. Neither pressing keys nor moving and clicking the mouse has any effect on the frozen game. The only action we can take is to push the reset button. The screen goes blank. This particular session of the game irretrievably disappears into an electronic graveyard, and the computer's checklist starts all over again.

I explain to the disappointed boys that the problem was likely caused by a programming glitch—possibly an error in the lengthy chain of instructions that tells the computer precisely what to do at every instant to create the sounds, images, and actions that make up the game. Given the number of programs dueling for space in our computer's active memory, it's also possible that the Sandiego program was simply trying to use a memory location—an "address"—that was already taken by another program operating in the background. Faced with such an unexpected roadblock, it had no alternative but to stop in its tracks.

Such faults have occurred often enough that both Eric and Kenneth are no longer caught by surprise when they happen. Nearly every one of the computer programs they use has glitches, though most of the defects aren't serious. The real wonder is that the system works as well as it does. The computer, together with its accessories, is by far the most complicated appliance in the house, and the computer programs that inhabit its memory add layer upon layer of complexity. Its astonishing capabilities range from speedy word processing to sophisticated drawing, and we can even listen to audio compact disks, played on the computer's CD-ROM drive, while using the writing or drawing programs.

However, when something goes wrong, it isn't easy to tell where or what the problem is. I know from bitter experience that there's usually no such thing as a simple fix. Repairs take time—often much more time than I intended to spend.

At *Science News*, the weekly newsmagazine where I work as a journalist reporting new developments in mathematics and physics, there are more computers. Some are in the open, identified by their screens and keyboards. Others lurk in the background, inconspicuously performing their assigned duties.

Consider the sleek, black telephone, bristling with buttons, that rests on my desk. It signals incoming calls and stored messages with diverse tones and flashes chosen from a suite of sounds and lights controlled by a bare-bones computer—a microprocessor—lodged somewhere in the instrument. It is also just one piece of a vast system of interconnected computers, telephone lines, and other hardware.

Such complexity makes pinpointing a fault a formidable task. On one occasion, soon after the installation of the new telephone system in our office, it proved extraordinarily difficult to make a long-distance call. After you dialed a number, you waited the usual six or so seconds for a ring or even the insistent beep of a busy line; unexpectedly the wait would stretch into ten, fifteen, twenty seconds of silence. Dialing the number again—and checking the telephone's liquid-crystal display to make sure the digits were correct—produced the same, frustrating result. Yet, occasionally, a call would get through.

Where was the problem? Had I forgotten a basic step by neglecting to read every detail of the instruction manual that accompanied the telephone? Was the microprocessor inside the telephone malfunctioning in some bizarre fashion? Was there a fault in the special-purpose computer, stowed away somewhere in my office building, that acts as the phone system's electronic switchboard? Was there a problem in one or more of the trunk lines that connected the office computer with the local telephone company's switching equipment or perhaps in that switching equipment itself? Could the difficulty lie within AT&T's computer-dependent communications network? Or was the fault lodged in the tangle of electronics and software at the other end of the line?

Despite, or perhaps because of, this maze of computers, cables, and other equipment, the nation's telephone system works exceedingly well. That's why it's easy to become dependent on the telephone. Any suggestion that the phones could be out of order for a few hours—or for an entire day—elicits near panic from the staff of journalists at *Science News,* as it would at nearly any other organization.

Computers are an inescapable facet of modern life. The general-purpose nature of the logic encapsulated in silicon integrated circuits allows programmed computers to operate in just about any field, to address an incredible array of tasks, and to supplant many human activities. Few businesses now function without computers of one kind or another. Microprocessors operate many of today's kitchen appliances, from microwave ovens to dishwashers. Automated systems manage traffic lights, supermarket inventories, and financial transactions, and they play crucial roles in such fields as air-traffic control and hospital management.

For one striking example of how rapidly this technology has transformed an industry, consider today's cars. Sophisticated, miniature computers now perform a variety of functions in cars. For instance, they can sense and automatically respond to mechanical problems and changes in driving conditions. The days when a kid could learn how to fix a car by tinkering with an old jalopy parked under an apple tree in the backyard are gone. Mechanics now have to go to computer school to learn how to decipher and diagnose the electronic circuitry and microprocessors that pervade today's cars.

Computers offer control, mastery—even a touch of magic. They contribute to a sense of liberation, a feeling shared by a diverse group of enthusiasts, ranging from computer mavens to doctors, lawyers, professors, musicians, artists, and writers. Children can readily lose themselves in an adventure game or in the joys of paint-

ing with an exotic electronic "brush." Engineers can construct and
deconstruct to their heart's content until their project looks and
works just right, without having to wait for prototypes from the machine or model shop. Freed from the finger-numbing drudgery of
typing draft after draft, writers can massage their words, sentences,
and paragraphs endlessly to create the perfect chapter or document.

So there's more to the appeal of computers than mere practi-
cality. Many people—though certainly not all—see computers as
sources of pleasure. They come to care passionately about what
they can do with a computer, especially in view of its apparent re-
sponsiveness to their needs. All that users require is the right software, and if it's not on the market, they can try to create it for
themselves. The opportunity is always there.

Frederick P. Brooks, Jr., a veteran computer scientist at the University of North Carolina, has described computer programs as messages from humans to machines. Programmers talk about the sheer
joy of fashioning complex, puzzlelike missives that precisely convey to the computer the instructions required to accomplish specified goals. In these messages, they create objects and
structures –incredibly sophisticated machinery—unconstrained
by physical reality. The only limit seems to be the cleverness of
the programmer.

Of course, there are definite limits to what one can accomplish
with software. Mathematicians and computer scientists have proven
that certain types of problems can't be solved by computers. In
other instances, solving a particular problem may require far more
time and data-storage capacity than anyone can ever provide. But
that still leaves lots of playing room.

"The programmed computer has all the fascination of the pin-
ball machine or the jukebox mechanism carried to the ultimate,"
Brooks says. "Programming . . . is fun because it gratifies creative
longings built deep within us and delights sensibilities we have in
common with all men."

Few media of creation are so flexible, so easy to polish and re-work, so readily capable of realizing grand conceptual schemes, he adds. However, unlike novels, musical compositions, or poems, computer programs are part of a cause-and-effect relationship. Mind is translated directly into action, a capability that is both exhila-rating and alarming.

An insistent silence enhances the magical effect. We give the command and no gears mesh and grind. No engine roars. No levers click-clack. Modern computation occurs silently, invisibly on a mi-croscopic scale, hidden away from prying eyes and ears. Neither the computer itself nor the stored programs—on floppy disks or magnetic tapes or integrated-circuit chips—have anything to say to the casual observer. Their secrets are locked away in a gray box or a secret code, and it takes special training and equipment to pen-etrate these mysteries.

Of course, the magic doesn't always work. An incorrect charac-ter or an out-of-place pause in the incantation can bring the whole system down. Human beings generally aren't used to maintaining in their other endeavors the high levels of perfection needed to create reliable software. Thus, programmed and used by humans, computers are far from infallible, and they hold a terrifying power to lose or destroy information. "A computer lets you make more mistakes faster than any other invention in human history," one commentator has wryly noted.

The evidence is everywhere, from misaddressed junk mail and errors in credit-card bills to temporary computer shutdowns that can paralyze business at travel agencies, banks, airports, and else-where. To make matters worse, in an increasingly competitive mar-ket, companies rush to get software products on the market, taking shortcuts and curtailing testing to save time. Indeed, commercial software producers seem to rely on their customers to do a lot of testing for them. So it's not surprising that commercial software usually contains errors and newly installed computer systems fre-quently fail. Instead of guarantees, commercial software packages invariably carry sweeping disclaimers, which in effect, say: "If this program doesn't work, tough luck!"

Imagine your reaction if you found that disclaimer on your car or kitchen appliance. What would happen to our society if everybody who wished to use a telephone, television set, car, detergent, or plastic toy were first obliged to learn at least a little about how it was made and how it works internally, and then to test it for hazards and other surprises? Why are software manufacturers allowed a sweeping disclaimer that no other manufacturer would dare to make?

Most people would be shocked and surprised if their new dining room table collapsed, but they accept as normal that, when they first install a computer system, it will fail frequently and become reliable only after a long sequence of revisions. The fact that practically no one expects software to work the way it should the first time out clearly demonstrates the formidable task experts face in ensuring that computer systems function properly and programs are free of errors—or bugs.

Obviously, there are many situations where we can't afford a single flawed performance. Computers that fly military or civilian aircraft, operate medical devices, manage railroad and other transportation systems, and perform crucial functions such as air-traffic control must work without fail. Yet, software defects and design errors have been implicated in plane crashes, telephone network failures, and fatal radiation overdoses. If computers have been at the heart of such mayhem, how can we ever trust them completely?

This book portrays the efforts of a handful of professionals dedicated to understanding and mitigating, if not eliminating, the myriad conditions that can lead to computer failures. Veterans of the computer age, these "bug hunters" have experienced both the fascination and the frustration of computation. They express awe at what computers have accomplished while tempering that wonder with an acute appreciation of the pervasiveness of human error—even folly—and the real limits of computing.

Often overlooked and largely underappreciated in both the computer community and the world at large, these men and women struggle to bring order to the anarchic practices still prevalent in computer system design and software engineering. Though they play crucial roles in industry, government, and academia, they also carry the taint that inevitably accompanies all bearers of bad news. They administer the necessary but bitter antidotes to software euphoria.

No one intends to build an unsafe system; no one plans to write a computer program that fails—sometimes with fatal results. But mistakes happen. They happen so often that computer professionals generally spend far more time fixing bugs than creating new programs. What distinguishes the individuals in this book is how seriously they take the occurrence of errors.

Achieving the necessary quality is no simple matter. Writing software is a quirky, labor-intensive scramble fraught with pitfalls and uncertainties. Imagine trying to write down step-by-step everything you do after getting up in the morning so that someone else could duplicate your actions. That's relatively easy if the steps are simple, the routine unchanging from day to day, and if the unexpected never happens. But few people live such well-ordered, predictable lives. Because minor details can have far-reaching consequences, events must be described with painstaking precision. It's also necessary to have appropriate responses to unlikely events. In the same way, a computer programmer must juggle myriad choices and contingencies while keeping clearly in mind the computer program's goals and functions.

Investigation of software errors and computer system defects is both a major preoccupation and a serious business. Although I use the terms "error," "fault," and "failure" interchangeably, the Institute of Electrical and Electronics Engineers (IEEE) has gone so far as to develop a standard that carefully defines and distinguishes these oft-used terms. Errors are defects in our thinking that occur when we try to understand information in order to solve problems or use particular methods. An error becomes a fault when it is written down—incorporated into software. A single error in reasoning may cause several different faults, or various errors may lead to identical faults. Failures arise when a software-driven system does something it's not supposed to do. This definition allows for the possibility that many different faults can sometimes lead to a particular type of failure, or that some faults may actually never cause a failure.

Such are the complexities of human error!

Nonetheless, computers usually perform their duties so invisibly and punctiliously that it's easy to take them for granted, and we generally accept without question all they have to offer. Few people ever doubt the accuracy of the numerical answer on a calculator display or contest the computer-tallied results of an election. The remarkable track record of these electronic servants has had a lulling effect, and these daily successes make it exceedingly tempting to continue expanding the realms over which they hold sway.

Nowadays, for example, no bank, stock exchange, or other financial institution can afford to operate without computers. Bankers and traders have access to spreadsheets, huge databases, and computerized models for analyzing the value and risks of complicated deals. And software vendors continue to offer new capabilities—new opportunities for financial manipulation—so that their customers can gain a potentially lucrative, split-second advantage over their rivals. As a result, banks and trading institutions are stepping into increasingly complicated financial ventures to create markets, to hedge their bets, and to design and manage all sorts of exotic loans and contracts.

What isn't clear is how reliable and stable the underlying computer programs used by such institutions are. And past history hasn't been particularly encouraging. In November 1985, the Bank of New York ended one of its trading sessions in government securities overdrawn by $32 billion—because of a software fault. To balance its books, the bank had to borrow $23.6 billion from the Federal Reserve Bank of New York, paying $5 million in interest on the loan. One result was a congressional investigation of the matter.

The problem itself stemmed from snafus and delays in changing the bank's software to increase the system's capacity for handling large volumes of securities. The programmers responsible for the software had rushed to complete their conversion but failed to make all the necessary changes in time for the following day's trading. The flawed software didn't work, and system managers found they could not return to the original program. There was no workable backup.

At various times, computer glitches have halted trading on the floors of a number of stock exchanges. In 1992, for instance, software problems at the Toronto Stock Exchange caused orders for certain shares to be garbled and entered at the wrong prices, leading to a massive disruption of the day's business. In the United States, the Securities and Exchange Commission in 1990 approved a policy calling for exchanges to assess the reliability of institutional computer systems on a regular basis, partly to call their attention to system vulnerabilities. Now, Wall Street brokerage firms are heading toward a 1995 deadline for switching to computerized, paperless trading of stocks and corporate bonds. There'll be no stock or bond certificates, no paper trail—just computer data and transactions that take place at lightning speed.

In 1994, the NASDAQ stock exchange, which has no trading floor and consists of a nationwide network of telephones and computers, suffered sporadic outages because of computer failures. One set of market-disrupting glitches resulted from the installation of new communications software. Another interruption occurred when a squirrel chewed through a cable that supplied power to the market's main computer installation, and backup power sources and computer systems failed to kick in promptly. These incidents proved embarrassing to NASDAQ, which bills itself as "the stock market for the next one hundred years."

And there may be more widespread disruptions in store as people rely increasingly on bank-issued debit cards to pay for goods and on automatic funds transfers to settle bills. Such "cashless" transactions—involving no currency or checks—now account for more than 18 percent of the $60 trillion that consumers, corporations, and governments spend annually.

In one case, in February 1994, automated teller machines (ATMs) at Chemical Bank in New York City mistakenly deducted a total of approximately $15 million from about a hundred thousand customer accounts. Until the problem was discovered, any customers making a withdrawal were charged double the withdrawal's actual amount on their accounts, although the printed transaction slip showed the correct amount. Only those people who later checked their balance—and knew what it should have been—realized there had been an error. The culprit proved to be a flawed instruction— a single line in an updated computer program the company had installed the day before the problem surfaced.

"There are similar episodes that take place all the time, but we never hear about them because the bank is able to get the accounts straight before it opens its doors in the morning," banking consultant Stuart R. Bloom commented in a *New York Times* report on the Chemical Bank incident. "The problem in this case is the ATM system is highly visible and runs twenty-four hours a day, seven days a week."

There are more than 85,000 ATMs in the United States, and they handle about eight billion transactions yearly. Their collective error rate is low, but not negligible. "Not enough people balance their accounts every month," Bloom added. "They assume that the bank is always right, and the bank usually is. But usually isn't always."

In a kind of perverse twist, human genius and human fallibility have joined to create an astonishing array of environments—from the patterns of logic etched into silicon integrated-circuit chips and the lines of instructions in a computer program to the enormous networks that bring different kinds of computers in widely distributed locations into precarious alliances—in which computer bugs can flourish. Hence, computer systems can suffer an amazingly diverse array of ailments.

Despite concerted efforts to prevent malfunctions and eliminate defects, problems continue to surface. Moreover, as computer designers and software engineers construct increasingly complicated systems, their chances of eradicating all possible bugs shrink to zero. And as techniques for countering problems and preventing failures grow more sophisticated, program developers seem to find new traps into which to stumble. Fuzzy thinking, combined with widely held misconceptions concerning the mathematics underlying computation, often lead them even further astray.

The steadily increasing speed of computers and the growing complexity of computer systems and networks make flaws ever more difficult to track down. Frequently, problems occur in environments where there are so many things happening simultaneously that by the time an error is detected, one no longer knows where or when it happened. Sometimes, an error lodged deep within a program can generate spurious information without ever being spotted.

As their craft gradually evolves into a discipline, practitioners of the software art find themselves in a capricious, volatile situation. Explosive technological growth has vastly outpaced the rate at which people can absorb the lessons of the past and formulate what they need to race into the future. In such a rapidly changing environment, the bug hunters spread their alarms not to stop progress but to manage it. Their caution flags are meant as signals to slow down, to be more careful, to take potential risks into account, and to heed past experience. They recognize, however, that some risk is inevitable and that a software solution—though imperfect—may offer greater safety than other alternatives.

Taking on ever greater responsibilities, computer systems also seem to be edging beyond human control and understanding. Designed to help us cope with complexity, the systems themselves are becoming too complicated for us to grasp in their entirety. This trend bodes ill for a future that could include unmanned oil tankers and other automated vehicles, automatically controlled, "smart" homes and office buildings, and the vast worldwide web of computers and communications equipment that is to serve as the information superhighway.

The fact that we can never be sure that a computer system will function flawlessly constitutes a fatal defect. It limits what we can hope to achieve by using computers as our servants and surrogates. As computer-controlled systems become more complex and thoroughly entwined in the fabric of our lives, their potential for costly, life-threatening failures keeps growing. Are we courting disaster by placing too much trust in computers to handle complexities that no one fully understands?

# FATAL DEFECT

# CHAPTER 1

# Inside Risks

---

**SLUMPING WITH AGE,** the rounded, heavily wooded peaks of the Vosges massif in northeastern France bear their long history with stolid fortitude. Castles, monasteries, and ancient fortifications guard the heights; medieval towns and patchworks of vineyards crouch in verdant valleys opening onto the plains of Alsace, adjacent to the Rhine River. In this part of France, summers tend toward the warm and sunny, though winters in the highlands can be markedly snowy and severe.

Near the northern tip of the Vosges, Mont Sainte-Odile rises to a height of nearly 2,500 feet above sea level. Pilgrims flock to the cluster of buildings that crown the peak to pay homage to Sainte Odile, who founded a convent on this site in the eighth century. Tourists trek the serpentine road to its summit for the magnificent view. On a clear day, they can glimpse the city of Strasbourg, about thirty-five miles to the northeast, and the Black Forest in neighboring Germany.

On January 20, 1992, a jet airliner plowed into a pine forest in the highlands near Mont Sainte-Odile. Just seven minutes away from the airport serving Strasbourg, the sleek Airbus A320 aircraft should have been flying at an altitude of nine thousand feet and descending slowly. Instead, it was at roughly twenty-four hundred feet when impact occurred at 7:45 in the evening. The airliner sent out no distress signal before the crash. It simply broke radio contact and abruptly vanished from radar. The nine passengers who survived the crash reported they had noticed nothing out of the ordinary until the moment of impact.

Eighty-seven people died in the crash. Darkness, freezing temperatures, thick fog, gusty winds, and deep snow hindered rescue efforts. It took more than two hours for a relief party to locate the wreckage and begin ministering to the survivors. By that time, news of the disaster was already spreading throughout the world by way of a vast electronic web of telephone lines, radio links, microwave transmitters, and the rest of the paraphernalia that constitutes our communal nervous system. As details of the crash emerged, radio and television reports, followed by newspaper articles, conveyed to a receptive audience the raw material to feed an age-old, ghoulish fascination with disaster and tragedy.

The next morning in Washington, D.C., James H. Paul, a staff member of the investigations and oversight subcommittee of the U.S. House of Representatives, noticed a report of the crash distributed by the Reuters news agency. The story bore the provocative headline: "Latest Crash Heightens Controversy over Airbus A320."

The A320's notoriety stems from the innovative technology incorporated in the airliner. Built by a consortium of French, German, British, and Spanish firms working together as Airbus Industrie, this narrow-bodied, twin-engine, 150-seat aircraft was the first of a new breed of airliner in which pilots control the airplane entirely through computers.

Paul's interest in the A320 crash arose in part from his role as one of the authors of a 1990 study called *Bugs in the Program: Problems in Federal Government Computer Software Development and Regulation* and in part from a concern he shared with others in the computer world about the difficulties of coping with the faults, or bugs, that plague just about any computer system.

Paul activated his computer's link with the Internet, a labyrinthine, worldwide network of computer networks and communications lines that can pass messages—electronic mail—from one computer user to another throughout the system. Using the address RISKS@CSL.SRI.COM, he sent out a copy of the Reuters article, giving his message the title: "Another A320 crash in France." Moments later, the missive arrived at a computer at SRI International, a high-technology research center in Menlo Park, California. There it remained stored until Peter G. Neumann, a principal scientist in the computer science laboratory at SRI, found a moment in his busy day to check the accumulated messages, which he fashions into the frequent reports of the Forum on Risks to the Public in Computers and Related Systems. Neumann made Paul's submission the first item in the report he distributed electronically on January 22, two days after the A320 crash.

**THOUGH NOVEL IN COMMERCIAL AVIATION,** the use of sophisticated computer systems to operate and keep high-performance jets in the air is essential for several types of military aircraft. The experimental X-29 military jet, with an unconventional, swept-forward wing design that makes it appear to be flying backward, is aerodynamically unstable, and no human pilot can fly it unaided. Only a computer can make adjustments quickly enough to keep the plane in the air. Already in service or being tested, the F-117A Stealth fighter, B-2 Stealth bomber, and F-22 advanced tactical fighter don't even have mechanical systems to act as backups for their computer controls.

The mechanics of steering an airplane and keeping it aloft were much simpler and more direct in the early days of powered flight. When Orville Wright coaxed the Wright Flyer into the air on December 17, 1903, near Kitty Hawk, North Carolina, his whole body came into play to control the plane. Lying prone beside the engine on the lower of a pair of cloth-draped wings, with his feet hooked over a spar, Wright used his arms to raise or lower the flying machine's horizontal front rudder to keep the plane at the proper angle. Cradled in a wooden yoke, his hips controlled wires that warped the wing tips for lateral balance. These hip movements also deflected the aircraft's vertical rear rudder to steer the plane.

The same fundamental principles underlie control of today's highly automated aircraft, but the link between a pilot and the consequences of his or her actions has become much less direct. Even with mechanical controls in a large, modern airliner, there's really no direct connection between the pilot's hand and the airplane's behavior. In a fly-by-wire arrangement, when a pilot pushes the plane's control stick to either side to turn the plane left or right, electronic sensors send signals to a set of flight-control computers. Instructions in the form of a computer program allow the computers to calculate what adjustments to make to the plane's rudder, flaps, or wing surfaces called ailerons. The computers, in turn, relay commands via an extensive network of electrical wiring to mechanical devices installed in the wings and tail to make the required changes in position of the movable parts. The plane then banks left or right. The pilot feels a comfortable amount of resistance in the control stick, just as if he or she were maneuvering the plane directly, but it's all a sham. A computer adjusts actuators in the stick to create the illusion of direct control.

In the Airbus A320, flight computers go so far as to tell the pilot how to fly the airplane, for example, by restricting how quickly or how far he or she can bank the airplane to the left or right. That's because the engineers who designed the aircraft gave the flight-control computers full authority to operate the airplane and override a pilot's actions when these actions force the plane into a situation that exceeds its capabilities and structural limitations. Faced with a sudden, wind-induced, stomach-churning drop in altitude, a pilot can react instantly by pulling all the way back on the control stick to nose the plane up, fully confident that the flight-control computers will ensure that the plane doesn't lurch into a deadly stall. At least, that's the theory.

Numerous pilots have complained that such delegation of authority to computers restricts their actions in an emergency. The pilots believe that a flight crew should be able to take any action deemed necessary to avoid crashing into a mountain or colliding with another aircraft—even if it means stressing the plane beyond its tolerances or flying it in an unconventional manner. Ironically, the computer-massaged A320 flies so smoothly and quietly under normal circumstances that both passengers and pilots have little sense that they are flying, and that in itself poses a risk of pilot overconfidence or complacency.

Reliance on computers for such critical functions as flight control also worries some computer professionals, whose experiences with a wide variety of computer systems have invariably demonstrated that no computer operates flawlessly all the time. Bev Littlewood, a professor of software engineering at City University in London, England, has long argued that because it's impossible to guarantee a sufficiently high level of reliability, the use of computers may be inappropriate for critical functions in aircraft, medical equipment, nuclear power plants, emergency communications systems, and other situations in which human life can be placed in jeopardy as a result of a computer failure.

**THE 1992 ACCIDENT,** the third A320 crash since the plane's debut in 1987, revived the controversy over automated flight control in commercial aviation, and it set off extensive discussion and debate in Peter Neumann's widely distributed electronic forum on the risks of computers. This popular forum has the immense appeal of juicy gossip bartered over a backyard fence, exchanged over coffee, or splashed across a tabloid newspaper's front page. Brought to light are the darkest secrets not of celebrities or neighbors but of computer systems and the people who design and use them. It's the kind of place where computer professionals can take secret delight in the misfortunes of others while remaining thankful that their own work has so far escaped such scrutiny.

The forum provides a window into a world normally hidden from outsiders. No bank wants to discuss publicly the travails of managing its computer-shuffled accounts. No financial institution is likely to admit openly that its software for calculating mortgage payments sometimes rounds numbers off incorrectly—often to the advantage of the institution. Few computer programmers willingly risk their jobs by complaining too loudly that the software for running a piece of medical equipment hasn't been tested adequately before the product went on the market. No advertisement is ever likely to display both the advantages of and the inevitable faults in a newly released version of a word-processing or personal-finance program.

It's easy to see why. People don't like to have their mistakes, oversights, or inadequacies hung out for everyone to see. Lawyers are ready to pounce, competitors are eager to take advantage, jobs are at stake, and national security concerns override candor. In an environment in which companies and individuals can afford to put forward only their best side—reinforced by good-news advertising and deft management of bad news—many errors and their perpetrators remain hidden in the shadows of expediency. But that makes the improvement of computer systems by learning from mistakes— a venerable strategy in engineering and other fields of human endeavor—much more difficult than it ought to be.

The Risks forum has become a place where a broad sampling of such errors is aired, furnishing lessons for all who care to pay attention. Imagine posting a notice on a bulletin board that interested individuals can peruse at leisure; they in turn can tack up their own scribbled notes, sometimes in response to items posted earlier. That's exactly what happens on an electronic bulletin board; only in this case, the users send messages to a central computer that displays the messages one by one in a long chain. It adds up to a curious kind of disembodied, multistranded conversation, stretched over time and space. Today's computer users share thousands of bulletin boards devoted to hundreds of topics and interests, accessible to anyone with a computer linked to a network or equipped with a modem to make a connection over a telephone line.

The Risks forum has one crucial feature that makes it less anarchic and better focused than many other electronic bulletin boards. It has a moderator—an activist master of ceremonies who scrutinizes the messages, selects which ones to pass on, adds comments (and in Neumann's case, a sprinkling of outrageous puns), and compiles them into digests that are then electronically disseminated. Issued as often as three or four times a week, each Risks forum digest runs to about a dozen typical book pages. And there's no shortage of material to keep these electronic pages filled week after week.

The tips come from a wide array of sources. Sometimes the items are posted anonymously. Sometimes they come from insiders with firsthand knowledge of a particular situation. More often, they are gleaned from newspaper accounts, magazine reports, and other media sources. Frequently, these items elicit replies, clarifications, comments, and contrary views from both knowledgeable experts and interested bystanders, who punctuate the debate with accounts of relevant personal experiences.

The idea of an on-line electronic forum devoted to the potential and actual risks posed by computer systems arose out of discussions among computer professionals belonging to the Association for Computing Machinery (ACM), which sponsors a variety of meetings and conferences, publishes numerous journals, and offers a means by which members interested in particular aspects of computing can pursue those concerns via special interest groups. On October 8, 1984, the council governing ACM passed a resolution that began:

> Contrary to the myth that computer systems are infallible, in fact computer systems can and do fail. Consequently, the reliability of computer-based systems cannot be taken for granted. This reality applies to all computer-based systems, but it is especially critical for systems whose failure would result in extreme risk to the public.

This resolution signaled a refreshingly candid and remarkably blunt admission of fallibility. Its second part, which also included a list of technical questions concerning the standards that any computer system ought to meet, stated that:

> While it is not possible to eliminate computer-based systems failure entirely, we believe that it is possible to reduce risks to the public to reasonable levels. To do so, system developers must better recognize and address the issues of reliability. The public has the right to require that systems are installed only after proper steps have been taken to assure reasonable levels of reliability.

The inauguration of the Risks forum in 1985 represented one response to these concerns, and Adele Goldberg, ACM president at the time, turned to Neumann, then chairman of ACM's Committee on Computers and Public Policy, to take charge.

NEUMANN HAD FIRST ENCOUNTERED COMPUTERS in 1953, while he was an undergraduate at Harvard University pursuing a degree in mathematics. In his junior year, he took a course in switching theory and hardware design, taught by applied mathematician

Howard Aiken. Aiken had been largely responsible for the development of a series of machines for automating calculation, and these giant electromechanical calculators had proven useful to the Navy during and after World War II for computing ballistics tables. It wasn't long before Neumann was convinced that his future lay with these marvelous but cantankerous machines.

Neumann spent the summer of 1953 as a fledgling programmer at the Naval Ordnance Laboratory in White Oak, Maryland, working with a large, starkly limited machine called the Card Programmed Calculator, which would nowadays appear woefully inadequate next to any modest electronic pocket calculator. In the following academic year, he wrote programs for the Harvard Mark IV, a descendent of the Mark I, the first computer designed by Aiken.

As a graduate student in applied mathematics, Neumann occasionally wrote programs for the Mark I, essentially a massive, general-purpose calculator that could carry out a sequence of operations specified by a program punched in a paper tape. By this time, the machine was more than a decade old and nearing the end of its useful life as a prodigious, chattering calculator of mathematical tables. But the chance of gaining firsthand experience with one of the famous, truly pioneering machines of early computation was hard to resist.

In the 1960s, while working at Bell Telephone Laboratories, Neumann became heavily involved in an innovative effort to develop a time-shared operating system for computers. To a considerable degree, computers of the 1940s and 1950s had been "personal computers." Only one person at a time could use them. As computers became faster and more sophisticated, programmers wrote special software to manage how and when a computer did its work. These supervisory programs—operating systems—could handle a number of other programs, taking individual jobs one at a time from a queue. Such "batch" systems represented a major step forward because several users could share a given computer. It was not uncommon to see users submit programs in the form of thick decks of punched cards to computer personnel, who would then stack the cards in a reader, awaiting the computer's pleasure. A few hours or a day later, the user could return to see what had happened.

The trouble with this system wasn't just that it was slow and inefficient. It also didn't allow the user to interact directly with the computer as it was executing a program's instructions. That made it extremely time-consuming to fix errors, test routines, or modify a program. With a time-shared operating system, a user could sit at any one of several terminals, all linked to the main computer, under the illusion that the entire computer was available to him or her. The operating system would juggle the needs of the various users' programs, sequentially allocating segments of time to each program in use.

Time-sharing technology originated with the Compatible Time Sharing System (CTSS), developed at the Massachusetts Institute of Technology in the early 1960s for the IBM 7090 computer. The Multiplexed Information and Computing Service (Multics) project, with which Neumann became involved, represented an ambitious attempt to improve on the original CTSS system—to do time-sharing "right," as Fernando Jose Corbató, who led the effort at MIT, has expressed it. Bell Telephone Laboratories and General Electric's computer department (later acquired by Honeywell) participated in this innovative, audacious venture, with GE providing computer hardware specially tuned to the needs of a time-sharing system.

For Neumann, the late 1960s was a hectic period, as he shuttled up to MIT for meetings every other week and managed the Bell Labs' share of the project. After four years of effort, however, Bell Labs ended its participation in 1969 and abandoned the system, though the remaining partners continued with the project.

In 1976, five years after he left Bell Labs and joined SRI, Neumann became editor of *Software Engineering Notes*, a quarterly publication put out by the ACM's special interest group on software engineering (SIGSOFT). Reflecting Neumann's resolute belief in the value of learning from past mistakes, several of the newsletter's pages were devoted to brief descriptions of examples in which malfunctioning computer systems had posed risks to human well being. Collected from colleagues, newspaper reports, and other sources, these juicy tidbits quickly became a popular, keenly read feature. To keep reminding people that such risks are very real, Neumann regularly compiled and distributed an index of what he calls "illustrative risks," drawn from the items reported in *Software Engineering Notes*.

**PRIVY TO SOME** of the deepest secrets of the computer netherworld but often enjoined from going public with such embarrassing tales because of promises of confidentiality in exchange for information, Neumann can at times no more than hint at the horrors he finds lurking in a world increasingly dependent on information processing and computer control. At other times, he can obtain no proof or the evidence is fragmentary. "Getting well-documented stories is exceedingly hard," Neumann has remarked. He takes particular pride in the several instances in which he was able to persuade a key, knowledgeable programmer or software engineer to write a thoughtful, instructive piece about a mishap, why it occurred, and what can be done to keep it from happening again.

One of the most famous cases involved preparations for the first space shuttle flight in 1981. On April 10, controllers called off the launch of space shuttle *Columbia* just twenty minutes before the scheduled liftoff. They were forced to confess before a huge, worldwide audience that a backup onboard computer would not work—presumably because of a fault somewhere in the computer system. It became "the bug heard round the world," and Neumann was able to persuade Jack Garman, who had helped develop the space shuttle software, to describe in detail what had happened.

To assure reliability, the space shuttle uses four identical flight-control computers, each simultaneously running the same program during shuttle launches and landings. A fifth, backup computer has a different set of instructions to do essentially the same job. Such an arrangement rests on the premise that the four primary computers can in some sense "vote" on what to do at each step. If the first pair of primary computers can't agree, then the second pair comes into play. If these two can't agree, the decision returns to the first pair. If that doesn't work, then the backup computer, which has a different program and shouldn't suffer the same fault as the primary computers, takes over.

This means that all five computers must be precisely synchronized, with their internal clocks agreeing within millionths of a second. Such clocks, based on counting the electrically stimulated vibrations of a sliver of quartz (similar to that found in a modern, battery-powered wristwatch) regulate computer operations. Like efficient, high-speed drill sergeants, they guarantee that a computer performs the appropriate instruction with each "tick," or clock cycle.

Shuttle engineers had encountered timing problems among the computers well before *Columbia*'s scheduled launch and had tried to fix them. But one of these repairs was itself flawed, and it introduced a 1-in-67 chance that when the system was switched on, the clocks of the primary computers would be one cycle ahead of the backup computer's internal clock. That's what happened when the shuttle's computers were turned on thirty hours before liftoff. Because the backup computer was programmed to ignore any out-of-sync signals, it could not receive the mistimed messages from the primary computers ordering it to proceed. The stalled computer waited and waited for a signal that could never come. Ironically, the redundancy introduced to increase reliability added to the system's complexity and to the probability of failure. Moreover, thousands of hours of testing had failed to uncover the bug.

"It was the kind of mistake that 'cannot happen' if one follows the rules of good software design and implementation," Garman noted in 1983. Yet, "it was the kind of mistake that can never be ruled out."

Several years later, on August 1, 1985, Neumann's first issue of the electronic Risks forum flashed onto computer screens. It offered a means of bringing such matters as the space shuttle incident to the attention of a much larger group than the software engineering community and the members of SIGSOFT. It also provided immediacy—real-time discussion of ongoing problems and concerns at a pace suited to the incredibly rapid evolution and application of computing technology.

Of all such problems, it was the controversy concerning the feasibility of developing and testing software to operate a national, space-based missile defense system designed to shield the United States from nuclear attack that took center stage in the inaugural issue. A key part of President Reagan's Strategic Defense Initiative (popularly known as "Star Wars"), the required system promised to be the most complex and most critical assemblage of computers and software ever contemplated. Indeed, the system's software would have to warn of an impending missile attack, pinpoint the attack's source, determine the likely targets, compute the missile trajectories, coordinate interception of the missiles, discriminate between decoys and warheads, and evaluate the effectiveness of each attempt to destroy a target—all in a matter of minutes.

ACM's leading journal had already featured letters and articles fiercely arguing the proposal's merits. Opponents pointed out its flaws and cited the inadequacy of known techniques for guaranteeing the system would work properly the first—and perhaps only—time it would be used. Proponents declared the problems enormously challenging but ultimately solvable.

Neumann focused on the resignation of David L. Parnas from a Department of Defense advisory panel consisting of nine computer scientists, who had been asked to evaluate the feasibility of developing computer systems for "battle management." In documents accompanying his letter of resignation, Parnas protested that computer science and software engineering offered no assurance that the immense difficulties posed by developing the required system could ever be overcome. In a devastating critique of the state of the art in computer science research, he argued: "Good software engineering is far from easy. Those who think that software designs will become easy [via new technologies] and that errors will disappear have not attacked substantial problems."

Four weeks after the inaugural issue, Neumann distributed its sequel, which featured varied reactions to items in the first issue and to the whole notion of an electronic Risks forum. In addition, it introduced several new topics, including problems with spacecraft software and with diagnostic aids that rely on intricate webs of rules to answer medical questions. Two days later, a third issue appeared, and the pace of publication has rarely slackened since, except for the brief intervals when Neumann has gone on vacation or when his own computer system has suffered a glitch.

In the early days, most of the contributors to the Risks forum were Neumann's friends or colleagues. But the number of contributors and the forum's audience grew rapidly. Nowadays, it's not unusual for Neumann to receive submissions from people he doesn't know at all, and his audience is so diverse and widespread that he has no real idea of its full extent.

BY JANUARY 22, 1992, the Risks forum had reached volume 13, number 5, and James Paul's submission on the A320 crash was just one of fifteen items mentioned in this issue. The four issues that followed all included comments on the A320 accident. Based largely on sketchy newspaper reports, these early, inconclusive speculations on what may have gone wrong turned on the resemblance between this disaster and two previous A320 crashes. In each instance, the pilot appeared to think that the aircraft was higher than in fact it was.

In June 1988, one of the first A320 models sold to Air France had brushed a patch of trees and crashed at the end of a demonstration flight at an air show. The commission of inquiry into the crash concluded there was nothing wrong with the airliner. In fact, the aircraft's computers had kept the plane's wings level to the end, preventing the plane from tumbling and making the accident worse. The commission's report accused the pilot of recklessness in flying too low. But the pilot, who survived the crash, insisted the airplane's equipment had failed to warn him of the loss of altitude just before the crash.

On a clear day in February 1990, an A320 operated by Indian Airlines was making its final approach for a landing at Bangalore airport when the airplane's speed dropped to a dangerously low level. Descending rapidly, the aircraft slammed heavily into a golf course just short of the runway. Ninety-two people died. Apparently, the pilot had inadvertently pushed the wrong button, setting the plane's engines on idle—a maneuver normally used for making a descent from higher altitudes but not immediately before a landing. Some critics of modern cockpit design have argued that the pilot had been lulled into a false sense of security. Following the accident, Airbus engineers altered the engine-regulating software to prevent the plane from flying more slowly than a certain speed.

Initially, the most puzzling aspect of the A320 crash near Mont Sainte-Odile was the pilots' apparent failure to realize that their airplane was descending rapidly and getting too close to the ground. Most airliners now have an alarm system that automatically calls out warnings in a loud, authoritative voice whenever the airplane descends too quickly or nears the ground. However, the system is mandatory only for international flights. Air Inter, the domestic French airline operating the A320 that crashed, had chosen not to install these alarms in all its planes. Company officials had complained that the alarms gave too many false warnings, and they noted that they were unnecessary anyway because the airline's pilots were quite familiar with the terrain across France.

In fact, when the alarm system had originally become mandatory nearly two decades ago, it was so unreliable that pilots quickly developed a hatred of its wrongly insistent voice, and many learned to ignore it. Subsequent improvements brought the error rate down considerably, but Air Inter believed the system was still too fallible. It's possible that even if the alarm had been functioning, the crew would have ignored it.

The A320 did have a separate system that uses a radio beam reflected from the ground to determine the aircraft's altitude and, in a simulated voice, calls out the reading at certain height intervals during maneuvers such as landing. It's the kind of task a copilot used to perform—and sometimes still does—to help the pilot, who must focus on flying the airplane. A recording of cockpit sounds during the last moments before the crash revealed that the pilots did get an automatic warning when the plane had dropped to just two hundred feet above the mountain's slope. But that left only one or two seconds before impact.

The question of how the pilots got into this untenable situation in the first place brought renewed attention to the A320's "glass cockpit." Instead of the round-dial instruments and toggle switches found in the cockpits of older planes, A320 pilots face an array of computer screens for monitoring the airplane and keyboards for typing in commands and making choices. Instead of dealing with separate dials and gauges for airspeed, altitude, rate of climb, and so on, they see the necessary information compactly displayed on their screens. But such a system also makes it more difficult for pilots to monitor long-term trends in airspeed and other flight parameters, so they must rely on automatic alerts or warnings to tell them about significant changes in the plane's status.

Computers handle so many functions that pilots spend a considerable amount of time typing in data, in effect programming the computers to set routes and prepare the plane for certain maneuvers. The computers automatically take into account such factors as fuel efficiency and aircraft safety, and they present the most economical and safest choices to the pilots. The crew can enter a flight plan, review it on the display, and command an efficiency factor. When coupled with the autopilot, the airplane's flight management system can control virtually the entire flight.

Indeed, increased automation has brought the pilot's role closer to that of a manager. Airplane designers have shown a strong preference for taking the pilot out of the control loop as much as possible. One manufacturer's preflight procedure simply requires that all lighted buttons be "punched out." The system isn't touched again except in abnormal situations.

In 1991, David Woods, a professor of industrial and systems engineering at Ohio State University, and his coworkers published an eye-opening study that highlighted the problems of such a high level of automation. The study concluded that automation actually makes flying more difficult for pilots, even for those with considerable experience in glass cockpits. Of the pilots surveyed, most had significant gaps in their knowledge of how computers run an airplane, and many admitted they had encountered situations in which they were surprised at what the plane did under certain circumstances.

Nearly all the pilots complained that the computer system was decidedly unfriendly, for example, flashing the notably opaque warning "invalid entry" instead of explaining why a certain typed input was unacceptable. Faced with complicated instructions and busy screens filled with information, pilots tended to adopt a cookbook mentality and use memorized "recipes" in their interactions with the computers. Even the option of having different ways of accomplishing the same task presented problems. Individual pilots developed alternative styles of programming, and that sometimes made it difficult for another pilot to step in and take over at a later time. Ironically, the system also gave pilots too little to do most of the time, which readily induced a dangerous complacency.

Reports by pilots to a clearinghouse on "incidents" involving aircraft have also presented evidence that the cockpit computers themselves sometimes function unpredictably and often display messages that don't make sense. It has almost become standard practice for pilots to ignore these apparently random and spurious messages by clearing the screen and, in some cases, by simply pressing a button to reset the computer at fault—in effect, turning it off for a moment. But because these incidents are rarely reported, it's not clear that anyone is looking into why a particular message would appear, and whether it may point to a defect in the software.

As news reports about the A320 crash in France dribbled out, contributors to the Risks forum freely offered their opinions, questions, and speculations, sometimes drawing on their personal experiences with aircraft or computers to offer minitutorials on their intricacies. The ensuing dialog helped clarify how the A320's maze of computers worked, how the A320 compared with other advanced commercial aircraft, and what kinds of training pilots should get when they become "system managers" in their cozy "glass cockpit" offices.

One issue that emerged early on was the possibility that the A320 pilots either had initially given the airplane's computer the wrong flight data or had selected an incorrect setting for establishing how fast the plane should descend in approaching the runway at Strasbourg airport. For instance, pilots could sometimes mistake the angle of descent, push the wrong button, and make the plane go down much faster than intended.

In the weeks following the crash, the Risks forum discussion of the A320 accident gradually tapered off, only to be revived briefly in March 1992, when the French government released an interim report on the crash. This report provided additional details, but it did not specifically apportion blame for the accident. It did lead to a strong recommendation that all French airliners—domestic and international—have functioning alarm systems to warn of excessively rapid descent and dangerous proximity to the ground.

A year later, the commission of inquiry into the crash rendered "pilot error" as its verdict—the same verdict reached in the two previous A320 crashes. The commission's report confirmed that the main problem was that the relatively inexperienced pilots had confused the "flight-path-angle" and "vertical-speed" modes of descent, either because they had forgotten to set the mode or had set it incorrectly. They didn't notice the error because of the similarity of the formats of the two displays. Other cues, such as altitude readings and vertical speed indicators on the main flight display, were apparently overlooked because the pilots were overloaded with various duties after last-minute changes in flight plan, some demanded by the Strasbourg control tower. They were presumably concentrating on the navigational display. The result was that the A320, descending at thirty-three hundred feet per minute instead of the usual eight hundred feet per minute, started at an altitude of five thousand feet and hit the mountain a minute later.

To some observers, the problem was less "pilot" and more "designer" error. If a badly designed system can overload a person with things to do and matters to consider, it seems unfair to blame the person for not keeping up. Ironically, increased automation, properly implemented, could help alleviate the problem. But it could also perpetuate procedures or systems that may be poorly thought out in the first place, merely adding to the complexity of flight management without solving the underlying problems.

Such lessons have not been lost on the Boeing Company. Sporting giant twin engines, each one wide enough to engulf a truck, the Boeing 777 made its debut in June 1994, rolling out of its hangar for a year of intensive, rigorous testing before going into service as the company's first "fly-by-wire" commercial airliner. Computers played a major role in the design, development, and construction of this aircraft. Interestingly, the company decided to go to "fly-by-wire" technology not for increased safety, reduced weight, or lower cost—though these were considerations—but primarily for greater ease of manufacture. Wires and electronic components are generally easier to install and maintain than their mechanical counterparts.

The design and building of the 777 also represented a new approach, one highly dependent on computers with special software to visualize exactly how the plane would fit together and what it would look like when it was finally built. With a digital design system, blueprints, drafting tables, large mockups, and bulky plaster models virtually disappeared from the plant floor. Using their elaborate computer system, engineers could readily check that various components would fit together snugly before the first piece of metal was cut. Viewing the brightly colored, three-dimensional images on the screens, they also identified a number of situations in which maintenance mechanics would have difficulties making repairs, and they made the necessary changes in the design based on this digital audition.

The 777's designers went to considerable effort to make the cockpit look and feel comfortable from the pilot's point of view. This was one reason why they decided to keep the pilot's control column, with its diminutive steering wheel, instead of going to the control stick used in the A320. Although the flight-control software does "protect" the airplane by keeping it flying within its specified tolerances, a pilot can override this protection feature by applying extra force on the steering wheel and column. Unlike the A320, the 777 also has a limited, mechanical backup system that would allow the pilot to fly the airplane manually if a computer failure occurs.

Of course, Boeing engineers and officials are doing everything they can to keep the 777 from becoming an item of discussion on Peter Neumann's Risks forum, though there are skeptics in both the aeronautical and computer communities who don't believe that the company can really prove that its new airliner's computer system has a sufficiently high level of reliability.

**THERE'S NO SHORTAGE** of tales of electronic woe for Neumann's digests or his occasional conference presentations. From 1987 through 1991, at a meeting devoted to "computer assurance" (COMPASS), held annually in Washington, D.C., Neumann made a practice of describing what he perceived as the worst computer-related risk of the year. In his 1991 address, he talked about weak links and correlated events—situations in which a few minor problems can add up to a huge disaster.

One of the examples Neumann cited was the February 25, 1991, failure of a Patriot missile battery to track and intercept an Iraqi-launched Scud missile, which subsequently struck a warehouse used as a barracks for U.S. forces in Dhahran, Saudi Arabia, during the 1991 Gulf War. The fault was traced to a 0.34-second error in the timing of a software-driven clock used for tracking the incoming missile. That error was large enough to prevent the system's radar from locking onto its target.

But several other factors exacerbated the problem. The original specifications for the Patriot were based on the assumption that the system would operate continuously for no more than fourteen hours before being shut down for maintenance. With that in mind, its computer programmers decided to use a set of equations for the tracking clock that happened to produce answers with an error of a millionth of a second per second. Such an error would normally pose no serious difficulties—if the system were shut down and reset after fourteen hours. But the crew operating the Dhahran battery didn't know about the accumulated clock error. By the time the Scud missile appeared, they had been running the system for a hundred hours without any discernible problems. By then, however, the clock error had grown large enough to make a difference, especially for tracking a missile traveling at several times the speed of sound.

Ironically, U.S. Army personnel had identified the timing problem a week earlier and produced a cassette incorporating the necessary corrections to the software. The new cassette, delivered first to Patriot batteries close to the Iraqi border, didn't arrive in Dhahran until after the Scud attack. Meanwhile, no one had warned the Patriot crew of the existence of the potentially serious fault in the tracking clock.

Neumann's comments at the 1991 conference were dramatically reinforced when local telephone service in Washington, D.C., and neighboring states was suddenly disrupted even as he was speaking. The problem apparently started in a single, faulty circuit board at a computer facility in Baltimore, one of four centers used by the Chesapeake & Potomac Telephone companies to route local calls. By itself, the circuit-board failure wasn't particularly serious, but it somehow triggered a catastrophic response in the computer software running the system. Nearly identical service disruptions also occurred on the same day in the Los Angeles area and a few days later in Pittsburgh and western Pennsylvania. In all three cases, the local telephone systems failed to cope with the effects of suddenly overloaded telephone lines. Flaws in the software controlling the sophisticated switches used for routing telephone calls and providing other telephone services proved to be the ultimate cause of this string of failures.

Programmers at DSC Communications Corporation, which had supplied the equipment and software involved in all the breakdowns, had inadvertently introduced the errors earlier in the year when they had made a few "minor" changes in the signaling software. At the time, company officials had deemed the alterations too insignificant to warrant the extensive testing usually conducted on new or revised software. Embedded in the software, the same bug surfaced wherever local telephone companies used this particular call-routing system.

**WITH HIS BEARD GRAYING** and his spectacles absentmindedly pushed up on his forehead, Neumann sometimes looks as if he carries the wearying burden of a computer-age Cassandra. As described in the *Iliad*, Cassandra was given the power of prophecy by Apollo but was cursed when the god decreed that no one would believe her. When facing disregard, Neumann sounds his most pessimistic and displays the drooping bearing of a man resigned to the pervasiveness of human folly, sometimes amused and sometimes saddened by the digital horrors he encounters. "We can't seem to learn the lessons well enough to avoid running into the same problems in the future," he said in 1991. "Things are getting worse faster."

And the Risks forum goes on. Recent contributions have ranged from personal accounts of the consequences of a telephone system failure at an airport and the hazards of using credit cards at automated gas pumps to comments on computer glitches involving banks, trains, and medical records, as initially reported in newspaper or magazine articles.

"The problems persist, either as new risks or incarnations of old ones, in just about every field that you can imagine," Neumann says. "More and more people are using computers implicitly or explicitly. There are many cases where we don't even realize that computers are controlling what's going on. As applications increase, we can expect the number of calamities to increase."

Immersed in the multifarious foibles of computation, Neumann remains dedicated to the notion that bringing computer problems to public attention will help improve software engineering and encourage greater care in implementing automation.

# CHAPTER 2

# Silent Death

**EVERYTHING APPEARED NORMAL** on the morning of March 21, 1986, when Ray Cox returned to the East Texas Cancer Center in Tyler to receive treatment for a tumor in his upper back. Several months earlier, doctors had removed a cancerous growth from this region of the thirty-three-year-old oil field worker's body, and the patient was now nearing the end of a regimen of therapeutic radiation treatments. Cox lay facedown on a table beneath the arm of a high-tech radiation therapy machine known as the Therac-25. Eight previous treatments had taught Cox that it was a painless procedure, no more disturbing than sitting for a photograph.

This time, however, Cox experienced a sharp jolt—a sensation resembling a strong electrical shock. At the same instant it shot through his body, he heard an unfamiliar buzzing sound from the equipment. His back felt as if someone had accidentally spilled a cup of scalding hot coffee over it. Alone in the radiation-therapy room, he started to pull himself from the treatment table. But just as he was getting up, a second burst struck his arm. Cox later recalled that it felt as if his arm had suffered an electrical shock and that his hand were leaving his body. Seeking help, he tumbled off the table, staggered across the room, and began pounding on the door.

Outside the room's seven-foot-thick concrete walls, the tech-

nician operating the computer-controlled radiation machine hadn't
seen Cox's reaction. On this particular day, the room's video monitor was disconnected and the intercom wasn't working. The only
indication that anything might be amiss was a cryptic message—
"malfunction 54"—that had appeared twice on the computer display outside the treatment room. Hearing the pounding, the
operator immediately opened the door. She was shocked to find a
shaken and injured Cox.

Cox was immediately taken to a nurse's station, where a physician examined him. Cox feared that he had suffered a radiation
overdose, but the Therac-25's computer display suggested, if anything, that an underdose had occurred. Showing reddened skin in
the treatment area but no obvious signs of serious injury, Cox was
sent home. The clinic's staff checked the machine but failed to uncover any problems. The Therac-25 went back into service the
same day and successfully completed its schedule of treatments.

That night, finding the pain in his neck and shoulder worsen-
ing, Cox checked into a hospital emergency room. A disfiguring
mass had developed under the skin on his back, and doctors suspected that he had suffered an intense electrical jolt. When the
cancer clinic was notified of this development, there was sufficient
concern about a possible electrical or radiation problem that clinic
personnel shut the machine down for testing. But they couldn't
reproduce the conditions that had led to malfunction 54. According to the manufacturer's manual, this particular error message
meant that the machine had delivered *either* an underdose or an
overdose of radiation, but there was no clear evidence this had happened. After an independent engineering report vouched for the
machine's electrical safety, the clinic returned its Therac-25 to service on April 7.

On April 11, malfunction 54 surfaced again at the Tyler clinic,
with the same machine and the same technician. The victim was
sixty-six-year-old bus driver Vernon Kidd, who was being treated
for a skin cancer on the side of his face. This time the intercom

was working, and the operator heard a loud noise and immediately rushed into the treatment room, where she found the patient moaning for help. Kidd had seen a brilliant flash of light, and he had heard an accompanying sizzling sound reminiscent of eggs frying. The side of his face felt as if it were on fire.

Three weeks later, Kidd died. An autopsy revealed a high-dose radiation injury to the right lobe of his brain and brain stem. Meanwhile, Cox lost the use of his left arm and experienced periodic bouts of nausea and vomiting. He was eventually hospitalized for radiation-induced damage to his spinal chord, which caused paralysis of both legs and other complications. He died in September.

On the same day as Kidd's overdose, the clinic physicist Fritz Hager mounted a concerted effort to determine what had gone wrong. He immediately shut the machine down and promptly notified its manufacturer, Atomic Energy of Canada, Limited (AECL), of the suspected overdoses. He began his own careful investigation, working diligently with the technician involved to try to re-create the conditions that had led to the deadly malfunction.

The machine at fault was the Therac-25 linear accelerator, a sophisticated, powerful device designed to fire a penetrating, high-energy beam of radiation deep into a patient's body to destroy embedded cancerous cells without injuring the surrounding tissue. From a port in its bulky, cantilevered arm, the Therac-25 could deliver radiation in two forms: either as a beam of electrons or as a beam of X rays. The accelerator produced the highly penetrating X rays by slamming a stream of high-energy electrons into a metal target. In the electron-producing mode, the machine would automatically move the metal target, decrease the electron beam's energy level, and send the beam directly to the tumor. Because a low-energy electron beam has less penetrating power than an X-ray beam, physicians use it to treat superficial cancers near the skin.

The crucial element in the machine's design was a turntable that carried the devices which modified the electron beam for a particular form of radiation treatment into position. At the first setting, high-energy electrons struck a metal target to produce X rays; in the second position, scanning magnets spread out the electron beam to a safe concentration for direct, electron-beam treatments; and in the third position, a mirror interrupted the electron beam path and a source of visible light illuminated the patient's body so the technician could aim it correctly. Three microswitches attached to the turntable monitored its position, and signals from these switches told the Therac-25's computer where the turntable was at any moment.

This computer had a variety of tasks that included monitoring the Therac-25's status, accepting input about the treatment desired, and readying the machine for the chosen treatment. After checking that the machine was set up properly, the computer turned on the electron beam in response to an operator command, then turned off the beam when the treatment was completed.

The computer could also shut off the beam if it detected a problem. Built into the software was a scheme designed to ensure that various sequences of operations occurred in the correct order at the right time. If it detected an electrical or mechanical malfunction, the computer could cut off power to the unit, either preventing the start of treatment or causing the machine to pause in the middle of a treatment. Depending on the fault, it would also provide a brief diagnostic message to alert the operator to the problem.

Working late into the night of April 11, Hager and the Therac-25 operator were finally able to elicit the fateful malfunction 54 condition. It turned out to depend on how quickly the technician typed in a particular sequence of commands. The operator had held her job for some time, and with experience, her data entry rate had increased greatly. She could rapidly type in the parameters for a given treatment and, if necessary, make corrections using editing features built into the Therac-25 software. In both the Cox and Kidd cases, she had noticed that she had accidentally typed in $x$ (for X ray) instead of $e$ (for electron) when selecting the treatment mode. This was a common error because most treatments involved X rays, and she was accustomed to typing $x$. The technician routinely fixed this error by using the keyboard's "up" key to move the cursor to the appropriate entry position on the screen, where she could correct the beam setting. Because all other parameter settings were correct, she hit the return key several times, leaving their values unaltered. When she reached the bottom of the screen, a message indicated that the parameters had been "verified" and the terminal displayed "beam ready."

The machine's software had to determine when data entry was complete and then send the commands necessary to set up the appropriate treatment. Because these operations required certain amounts of time, the program had to juggle the timing of various operations to keep them properly in sync. But it turned out that flaws in how the software was written allowed moments of vulnerability, and there were no checks built in to detect such inconsistencies.

In the Cox and Kidd cases, the computer had already noted that the operator had reached the end of the screen and had requested the X-ray mode. It proceeded to order the setting of the magnets required for generating the high-energy electron beam needed to create X rays. That operation requires eight seconds. The operator, however, edited her screen so quickly that she completed the resetting of the parameters in less than eight seconds—a time interval during which the computer was also waiting for the completion of magnet setting. Although the screen reflected the changes, a key part of the program required for configuring the machine's beams did not. That misstep permitted an unobstructed, high-energy electron beam, capable of destroying any tissue in its path, to escape the accelerator's radiation port.

In essence, an overdose occurred whenever the radiation prescription data were edited at a sufficiently rapid pace. With practice, Hager eventually could repeat the procedure quickly enough to elicit malfunction 54 whenever he wanted. At first, AECL engineers themselves couldn't reproduce the error. They finally got a similar malfunction on their own machine after Hager told them of the crucial role that timing played in the problem. Subsequent testing at AECL revealed that under these conditions, instead of delivering a relatively safe level of radiation, the Therac-25 would mistakenly administer a dose at least a hundred times more potent than intended.

On April 15, AECL submitted an accident report to the Food and Drug Administration (FDA). On the same date, it sent a letter to each of the eleven clinics using a Therac-25. The letter recommended a temporary "fix," stating:

> Effective immediately, and until further notice, the key used for moving the cursor back through the prescription sequence (i.e., cursor "UP" inscribed with an upward pointing arrow) must not be used for editing or any other purpose.
>
> To avoid accidental use of this key, the key cap must be removed and the switch contacts fixed in the open position with electrical tape or other insulating material. . . .
>
> Disabling this key means that if any prescription data entered [are] incorrect then [an] "R" reset command must be used and the whole prescription reentered.

This limited action didn't satisfy the FDA, and on May 2, 1986, the agency declared the Therac-25 defective. FDA officials also complained that the AECL's letter to the machine's users failed to describe adequately the defect and the hazards associated with it. "The letter does not provide any reason for disabling the cursor key and the tone is not commensurate with the urgency of doing so," the FDA wrote. "In fact, the letter implies the inconvenience to operators outweighs the need to disable the key."

Ironically, complaints from operators that data entry for an earlier model was too laborious had led the manufacturer to rewrite parts of the original computer program to make the machine more user-friendly. The new procedure allowed an operator the option of making simple corrections without retyping lengthy strings of commands. Nonetheless, by making these changes, the manufacturer had inadvertently introduced a fatal defect.

But there was much more to the story than an isolated software defect. The Texas incidents turned out to be just two of six such cases of accidental overdoses, and it would eventually become clear that attributing all of these accidents to a single cause was mistaken.

WHEN THE FAMILIES of the victims of the Texas accidents filed a lawsuit against AECL, the cancer center, and the supervising physician, the prosecutors brought in Nancy G. Leveson as an expert witness. A professor of computer science then at the University of California in Irvine, Leveson had played a key role in pioneering the notion of software safety. This endeavor represented the extension to computer systems of techniques conventionally used for anticipating and avoiding accidents in industrial and other settings. By specifically identifying and paying attention to conditions that lead to accidents, engineers could then eliminate these hazards or reduce them to an acceptable level.

Consider the example of a traffic light, which must repeatedly cycle through green, amber, and red. Such a light can fail in a variety of ways, ranging from a lightbulb burning out to a complete power failure. Its circuitry may also fail in such a way that the light stays green or red. A traffic light stuck at red or green presents a hazard because a driver doesn't necessarily realize the light is broken. Traffic engineers have designed systems that when they fail, go to a flashing yellow or flashing red light, to indicate clearly that drivers must proceed cautiously. Of course, this is to no avail if a complete power failure occurs. At critical intersections, the activation of an emergency power supply may provide at least part of the answer.

Bringing this type of approach to computer systems signified a considerable departure from conventional practice in software engineering and computer systems development. Computers themselves are relatively safe devices. They rarely explode, catch fire, or cause physical harm. At the same time, computer programmers have generally taken the approach of aiming to write perfect software, by refining it through testing and other techniques until all bugs are eradicated. Leveson's radical notion was that because it is virtually impossible to achieve the required perfection, any system should be designed so that whatever the software does or doesn't do, there is no significant threat to human life.

Engineers who built robots, bridges, or integrated-circuit chips don't have the luxury of making their products—their hardware— perfect because they know that a physical system always has a chance of failing. "So they build in protection against these failures—against various kinds of hazards," Leveson says. "That's what I've done in software. We have to build software to be safe. Just trying to be correct is not enough."

She goes on, "There's nothing wrong with making software as reliable as possible, but that's not going to make it safe. Programmers will write the code [lines of a computer program] to match the requirements, but the requirements may not be right. Most accidents occur because the requirements are wrong or incomplete, not because of a coding error."

In a 1991 paper outlining techniques for evaluating and achieving software safety, she wrote:

> Computers can contribute substantially to accidents when they operate as a subsystem within a potentially dangerous system. Examples include computers that monitor and control nuclear power plants, aircraft and other means of transportation, medical devices, manufacturing processes, and aerospace and defense systems. Because computers are not unsafe when considered in isolation, and only indirectly contribute to accidents, any solution to computer or software safety problems

must stem from and be evaluated and applied within the context of system safety.

Leveson had already established a strong reputation in this specialty by the time the Therac-25 accidents occurred. She had majored in mathematics at the University of California at Los Angeles but discovered that her only choices for employment were teaching (her mother's preference, which meant that it couldn't be hers) or actuarial work, which sounded too boring. She settled on computer science as a possible career, and because there was no computer science department yet, ended up at the School of Management, where a psychologist was teaching a course in artificial intelligence. Leveson's first experience with programming computers was sufficiently interesting to keep her going, and she emerged in 1967 with a master's degree in computer science and operations research. Her first job was as a systems engineer for IBM, providing technical support for owners of IBM equipment.

But in the spirit of the late 1960s and early 1970s, Leveson dropped out, joining the legions of flower children and vagabonds roaming the country and the world. "I've always been interested in cultural anthropology, so I spent time in places like the highlands of New Guinea with basically stone-age people," she says. "I also spent a fair amount of time in Asia, staying with families who I met on trains, living in Salvation Army hostels, and sleeping on beaches."

Two years of this wandering was enough to get Leveson to start thinking about doing something with her life, something more than simply collecting a paycheck. She taught high school mathematics for a year, then decided to return to school to learn to work with emotionally disturbed children. She ended up working toward a doctorate in cognitive psychology, but for a variety of reasons drifted back into computer science. In 1980, she obtained her doctorate in computer science from UCLA.

As she settled down to her career, Leveson maintained a practical, down-to-earth orientation geared to her earlier experience with real-world computer systems and especially with people. Her subsequent focus on the safety side of software engineering reflects a latent sense of idealism, a concern that she should be spending her life doing something positive for the world.

It's a role Leveson likes. As one of the few women playing a prominent role in a discipline overwhelmingly dominated by men, she tackles not only computer problems but also the ingrained attitudes that conspire against women playing a more important role in the computer industry.

Drawn into the Texas case, Leveson had a unique opportunity to learn from the inside—to be present at and contribute to an autopsy of a software and systems failure. But the case was settled out of court, and all documents and files gathered as evidence were sealed, shut away from public view. Several other cases involving victims elsewhere were similarly settled out of court.

Though she was barred from talking about the specific details of the Texas case, this didn't end Leveson's interest in documenting the Therac-25 tragedy as a case study in software safety—or rather the lack of it. There were alternative sources of information, especially government records made available through the Freedom of Information Act. Leveson joined forces with an attorney, Clark S. Turner, who had returned to school to obtain a doctorate in computer science. Admitted to the bar in California, New York, and Massachusetts, he was now shifting his focus to technological issues.

Together, Leveson and Turner spent nearly three years collecting information. Obtaining public documents from the Food and Drug Administration and various agencies in Canada, they amassed a huge volume of paper. From these materials, they reconstructed the Therac-25 accidents, even though a number of pieces belonging to the puzzle were unobtainable.

Leveson and Turner outlined their basic premise at the beginning of their report, which finally appeared in 1993. "Our goal is to help others learn from this experience, not to criticize the equipment's manufacturer or anyone else," the authors declared. "The mistakes that were made are not unique to this manufacturer but are, unfortunately, fairly common in other safety-critical systems."

This investigation also served as a vehicle for demonstrating the validity of Leveson's notion that accidents are rarely simple events. Too often, she contends, complicated chains of events are attributed to a single cause. It's this tendency to single out particular causes that makes Leveson critical of many of the reports filed in the Risks forum described in Chapter 1. "Most accidents are systems accidents; that is, they stem from complex interactions between various components and activities," she says. "To attribute a single cause to an accident is usually a serious mistake."

And so it proved in the Therac-25 case.

THE THERAC-25 was a direct descendant of the pioneering Therac-6 therapeutic linear accelerator, jointly developed in the early 1970s by AECL and the French company Thomson CGR. The original machine produced only X rays. It had a computer, used for a few housekeeping chores, which was not essential for running the machine itself. The Therac 6's success led to the more sophisticated Therac-20, which could produce beams of both X rays and electrons. Its computer played a greater role in the machine's operation, but the software did not directly control the safety features built into the Therac-20. Electrical circuitry incorporating fuses and circuit breakers automatically cut off the beam if the machine was operated incorrectly or unsafely.

The relationship between AECL and Thomson ended in 1981. Meanwhile, AECL had developed the Therac-25, a showcase product more compact and versatile than its predecessors. The new unit was also easier to use, the manufacturer advertised. Controlled by a minicomputer, this machine relied on software to operate it and ensure safety. The logic of specific chains of instructions in the program replaced the electrical circuitry and other hardware that had previously guaranteed that such a machine had proper safeguards.

The Therac-25's computer program, consisting of about twenty thousand instructions, had been written by a single programmer over a period of several years. It incorporated parts of the Therac-6 and Therac-20 programs, along with a great deal of new material tailored to the Therac-25's special features. Curiously, very little information about who this individual programmer emerged from the Texas trial. AECL could furnish no employment records, and company employees who filed depositions could provide no information about this person's education, qualifications, or experience. It is known that this programmer left the company in 1986.

Leveson and others who had a chance to examine the software were appalled by the mess they found. There was very little documentation—nothing written out to explain in plain English what different parts of the program did. There was no analysis demonstrating that key strings of instructions led to appropriately timed actions. There was no evidence that the software itself had been extensively tested before being bundled with the machine. The whole package displayed shoddy, naïve programming, but unfortunately, it typified the informal, undisciplined approach taken by many software developers in the 1970s.

Only eleven Therac-25 machines were in use when its problems became known. There were five at sites in the United States and six in Canada. Most of them had been installed in 1983, and they had apparently functioned without incident for some time.

The first known accident occurred at the Kennestone Regional Oncology Center in Marietta, Georgia, on June 3, 1985. Following removal of a malignant breast tumor, a sixty-one-year-old woman was receiving follow-up radiation treatments. On this occasion, the patient felt a tremendous blast of heat when the machine was turned on. She promptly complained to the technician operating the machine that she had been burned. The technician replied that this was impossible. Nonetheless, although there were no marks on her body, the treatment area did feel warm to the touch. Red marks accompanied by swelling appeared later that day on the patient's body, and eventually the pain became intense enough to keep her from moving her shoulder. It was only many months later that it became clear that she had suffered a radiation overdose.

Of course, there were suspicions right after the incident that a radiation overdose might have occurred. Kennestone physicist Tim Still called AECL personnel to ask whether the Therac-25 could operate in its electron mode without scanning to spread out the beam. Three days later, engineers at AECL replied that improper scanning was impossible. There was no serious investigation of the accident, and the company never informed other Therac-25 users of the problem at Kennestone. The patient's lawsuit against the hospital and AECL was settled out of court. The woman continued to suffer from intense pain due to the injury, and one of her breasts had to be removed. She died a few years later in a car accident.

The second in this series of accidents occurred on July 26, 1985, at the Ontario Cancer Foundation clinic in Hamilton, Ontario. A forty-year-old patient was receiving her twenty-fourth Therac-25 treatment. The operator activated the machine, but after five seconds, it shut down. The computer screen showed the error message "H-tilt," indicated that no dose had been administered, and signaled a "treatment pause." Noting the nil dose and the pause, the technician went ahead according to standard operating procedure with a second attempt at the treatment, pressing the "P" key (the "proceed" command). The same thing happened again, and again, and again. Finally, after the fifth pause, the machine shut itself down. A hospital service technician called in to check the machine could find nothing wrong. But the patient complained of a burning sensation, which she described as an "electric tingling shock" in the treatment area in her hip. Within three days, she was experiencing severe pains in her hip, and the affected region had reddened and become swollen. Suspecting a radiation overdose, clinic staff shut down the Therac-25.

The patient died on November 3, 1985. An autopsy revealed the cause of death as a particularly virulent form of cancer, but had the woman not died, she would have required a total hip replacement because of the radiation damage.

AECL investigated the accident, but it couldn't reproduce the malfunction that had occurred. The manufacturer suspected that one of the microswitches indicating the turntable's position had suffered a transient failure. Indeed, in following up this line of investigation, AECL engineers discovered some weaknesses in their design and uncovered some potential mechanical problems that could improperly position the turntable. AECL ended up informing Therac-25 users that, until further notice, they ought to visually confirm the turntable's position before beginning a treatment. Three months later, AECL came up with an improved system, and it altered the software to keep better track of the status of each of the microswitches and the turntable position. Nonetheless, AECL could not say with certainty that this had been the actual cause of the Hamilton accident.

The next serious malfunction occurred in the Therac-25 at the Yakima Valley Memorial Hospital in the state of Washington. This machine had been modified a few months earlier to meet the new requirements promulgated as a result of the Hamilton accident. After a treatment on December 11, 1985, a woman developed a peculiar reddening of the skin on her right hip. This discoloration showed up as a distinctive pattern of parallel stripes, which appeared to match a set of slots on a part of the Therac-25's beam-delivery system. Despite this reaction, the patient continued with her treatments into January, while the clinic staff monitored the skin condition and tried to figure out its cause. But they could come to no firm conclusion. Reassured by the manufacturer, they had no reason to believe the patient had suffered a radiation overdose.

The patient survived, suffering minor disability and scarring related to the injury.

No alarm bells had yet sounded at AECL. It was the pair of Texas accidents a few months later that finally pinpointed a serious software defect and precipitated remedial action by the users and at AECL, which had been stubbornly blind to the possibility of a software fault.

By this time, users of the various kinds of Theracs had taken a more active role in diagnosing their own machines, communicating their findings and fixes. When physicist Frank Borger of the University of Chicago Joint Center for Radiation Therapy heard about the Therac-25 software problem, he decided to find out whether the same thing could happen with the older Therac-20. He suspected that the AECL programmer had borrowed certain software routines from the Therac-20, and he felt it was worthwhile looking for a software timing fault in the older machine.

At first, Borger was unable to reproduce the error on the university's machine, which was used to teach students. But then he remembered an unusual pattern of activity associated with the device. Whenever a new class of students started using the Therac-20, the unit would often shut down because of the tripping of various fuses and circuit breakers. Such failures occurred as often as three times a week while new students operated the machine, but then the shutdowns would disappear for months, until a new crew of students started using it.

Borger's interpretation of this distinctive pattern was that new students make all kinds of typing errors and quickly develop "creative methods of editing" the parameter values shown on the screen. He determined that certain editing sequences were closely correlated with blown fuses. In fact, the underlying timing fault responsible for the stoppages was identical to that in the Therac-25 software. The crucial difference in the case of the Therac-20 was that this machine's protective circuits independently monitored the machine's status, and if something was amiss, a fuse would blow, shutting down the machine so it would be unable to deliver a radiation overdose. Thus, the software error that led to a fatal result with the Therac-25 was merely a nuisance in the Therac-20.

"A lesson to be learned from the Therac-25 story is that focusing on particular software bugs is not the way to make a safe system," Leveson and Turner conclude. "The basic mistakes here involved poor software engineering practices and building a machine that relies on the software for safe operation. Furthermore, the particular coding error is not as important as the general unsafe design of the software overall."

The last incident in this sequence of tragic errors occurred on January 17, 1987, at the Yakima Valley Memorial Hospital. This time, another patient was overdosed, but an entirely different software error appeared to be the cause. The result, nonetheless, was the same. The machine accidentally exposed a patient to an intense electron beam. The victim died in April of that year from complications related to the overdose.

By this time, AECL was taking the whole matter seriously enough to develop a plan for correcting a wide range of machine faults that had contributed to these accidents or had the potential for causing accidents in the future. In a *Wall Street Journal* article on January 28, 1987, Walter Downs, a quality-assurance manager at AECL, admitted, "There are too many combinations of features to guarantee that the [radiation] beam won't come on too intensely."

The company also brought in outside consultants to analyze the machine's software thoroughly, and this review led to a number of software changes. "I'm convinced the software has more bugs," Leveson says. "But efforts to make the machine safe overall, despite software bugs, have decreased the risk of an accident occurring because of the software."

Ed Miller of the Center for Devices and Radiological Health at the FDA expressed a similar sentiment. "FDA has performed an extensive review of the Therac-25 software and hardware safety systems," he reported in 1987. "We cannot say with absolute certainty that all software problems that might result in improper doses have been found and eliminated. However, we are confident that the hardware and safety features recently added will prevent future catastrophic consequences of failure."

He added, "I, for one, will not be surprised if other software errors appear with this or other equipment in the future."

The advent of computer-controlled linear accelerators nearly two decades ago promised greater convenience, flexibility, and speed for radiation therapy, which remains a basically safe, routine procedure that has saved the lives of many people with cancer. Indeed, the Therac-25 itself has operated successfully thousands of times, providing incalculable benefits. But on six occasions, it behaved unpredictably, and it unexpectedly betrayed its human masters—and the patients it was supposed to help heal.

The Therac-25 problem could have been avoided. A proper safety analysis would have identified the missing target as a potentially dangerous situation, and the machine could have been programmed so that it could not operate in the high-energy electron mode without the metal target or spreading magnets in place. It's also possible that such a complex, computer-driven machine wasn't even necessary; by sacrificing a little convenience and flexibility, a machine with a simple on-off switch and a timer could probably have done the same job—with a much smaller chance of failure. In fact, similar machines manufactured by other companies have usually retained mechanical controls to ensure safety, and by and large, these systems have a good record.

Nonetheless, medical physicist J. A. Rawlinson has remarked, "In the past decade or two, the medical accelerator industry has become perhaps a little complacent about safety. We have assumed that the manufacturers have all kinds of safety design experience since they've been in business a long time. We know that there are many safety codes, guides, and regulations to guide them, and we have been reassured by the hitherto excellent record of these machines. . . . Perhaps . . . we have been spoiled by this success."

At the same time, the convenience, apparent flexibility, and exquisite power of software control has a seductive appeal. Hospitals and other customers are invariably much more willing to pay extra for added features than for the relative safety that simplicity offers. "Often, it takes an accident to alert people to the dangers in technology," Leveson and Turner conclude.

"Accidents are seldom simple—they usually involve a complex web of interacting events with multiple contributing technical, human, and organizational factors," they continue. "One of the serious mistakes that led to the multiple Therac-25 accidents was the tendency to believe that the cause of the accident had been determined (for example, a microswitch failure in the Hamilton accident) without adequate evidence to come to this conclusion and without looking at all the possible contributing factors. Another mistake was the assumption that fixing a particular error (eliminating the current software bug) would prevent future accidents. There is always another software bug."

Yet, "it is nearly useless to ascribe the cause of an accident to a computer error or a software error," Leveson and Turner note. "Certainly software was involved in the Therac-25 accidents, but it was only one contributing factor. If we assign software error as *the* cause of the Therac-25 accidents, we are forced to conclude that the only way to prevent such accidents in the future is to build perfect software that will never behave in an unexpected or undesired way under any circumstances (which is clearly impossible) or not to use software at all in these types of systems. Both conclusions are overly pessimistic."

**THE PROBLEM** is certainly not an isolated one. In 1985, M. Frank Houston, then at the FDA and now a consultant, warned: "A significant amount of software for life-critical systems comes from small firms, especially in the medical device industry; firms that fit the profile of those resistant or uninformed of the principles of either system safety or software engineering."

In some cases, the software is literally the product of a single programmer, working in a basement or a garage. Operating on shoestring budgets, the companies involved are often too small to absorb the cost of extensive testing and too focused on the rush to get a product on the market. Inevitably, they fail to anticipate certain situations that the particular device may encounter.

The early 1980s, for example, saw a number of cases involving faults in the design and testing of software for microprocessor-controlled heart pacemakers. In one case, an external device for programming a pacemaker set it at the wrong rate, and the patient died. The manufacturer recalled the device and corrected what it believed was the most likely sequence of commands that caused the error. However, soon after the company had tested the change and released the corrected version, field personnel alerted the company that there was also another way to produce the same mistake. The company had to start all over again.

"Programmers and managers in small businesses, many building safety-critical software, perceive extra work in software engineering," Houston noted. "Managers, in particular, tend to see how easily software can be modified instead of the problems that hastily built software manifests after delivery. On the average, computer science education propagates outmoded habits of software design and programming, which students carry into their careers, although there are more enlightened curricula at some universities. Moreover, many experienced programmers have not kept current with the latest techniques of software engineering. Programmers and their managers habitually discount the importance of early design and planning phases of a project, preferring to get right to prototypes and coding, which is easier to formulate and evaluate than design."

Paradoxically, the situation has both improved and gotten worse. Physicians and hospital personnel using particular types of equipment and techniques have established strong user groups to compare experiences and alert one another to equipment vulnerabilities via newsletters, conference sessions, and electronic bulletin boards. User groups can put considerable pressure on companies manufacturing and marketing products that are defective and don't meet needs, even to the extent of promulgating standards. In fact, it was the activity of such a group after the Therac-25 accidents in Texas that helped identify the crucial problems and forced the manufacturer to pay attention and step up its investigations.

One of AECL's problems was its poor handling of hazard reports from customers, which made it needlessly difficult for the company to identify serious deficiencies in its product. The company's awareness of possible defects built up unconscionably slowly. A chastened AECL is still in business, and its radiation therapy equipment division now operates independently under a new name, Theratronics.

Fortunately, more companies are using better software engineering practices, from the insistence on complete documentation to careful planning before coding begins and the use of approved, well-established techniques for writing software. But there's still a long way to go. With these improvements come the problems of coping with increased complexity. This increase in complexity wipes out many of the gains made by using better programming techniques.

"Furthermore, these problems are not limited to the medical industry," Leveson and Turner contend. "It is still a common belief that any good engineer can build software, regardless of whether he or she is trained in state-of-the-art software-engineering procedures. Many companies building safety-critical software are not using proper procedures from a software-engineering and safety-engineering perspective."

Moreover, the frequent lack of communication between the designers and engineers building a system and its users continues to pose a variety of problems. Computer-based medical equipment generally performs just as it was designed to perform. The trouble is that its users may not be aware of or understand the designer's intent, especially when the device does something that the user doesn't intuitively expect.

Take the example of a heart and lung machine used for life support during open-heart surgery. Controlled by a computer, the machine may occasionally stop suddenly and without warning for a variety of reasons specified by its designers. But the reason for a particular shutdown may not always be obvious to the perfusionist charged with keeping the patient alive, and different people react in different ways to an intentional delay that looks like a crisis. Some trust the machine and wait for additional information, while others hurriedly press every button in sight to try to get the machine to respond.

Of course, the system's designers anticipated that someone might inadvertently activate a switch, so they built a time delay into the machine's switches so that even if the correct button is pressed first, there may be a distinct pause before the machine reacts. But that feature may cause a technician's panic to escalate to sheer terror when the machine doesn't react promptly.

This was also true of the Therac-25. In clinical use, the machine typically paused or shut down so often—leaving a trail of forty or more incredibly cryptic and uninformative diagnostic messages per day—that operators quickly learned to ignore the warnings. Because the majority of these messages indicated nothing more serious than that the beam intensity was slightly less than it should have been, they simply learned to respond to all errors by pressing the "P" ("proceed") key. In fact, there were so many "safety" checks that the machine's interface actually encouraged operators to run the Therac-25 in a hazardous fashion.

"An operator involved in an overdose accident testified that she had become insensitive to machine malfunctions," Leveson and Turner state in their report. "Malfunction messages were commonplace—most did not involve patient safety. Service technicians would fix the problems or the hospital physicist would realign the machine and make it operable again. . . . The operator further testified that during instruction she had been taught that there were 'so many safety mechanisms' that she understood it was virtually impossible to overdose a patient."

At the same time, although regulatory agencies are asking tougher questions, they can't readily check that a manufacturer's claims are truly representative of the product in question. How can the FDA verify that a particular computer program does what it's supposed to do—nothing more, nothing less? Furthermore, computer programs are easy to change and can be used in many different ways. If corrections are made and new features added, how much scrutiny should the modified version of a previously approved computer product undergo? As experience with software for other applications has shown, the task of checking software quality can be overwhelming.

CENTRAL TO LEVESON'S RESEARCH is the notion that all human activity involves some degree of risk. There's always a chance that a traffic light will fail, an airplane will crash, or a bridge will collapse. Indeed, technological progress itself requires some risk, whenever a new, unproved technology is introduced—whether in the form of new materials for airplane fuselages or novel control strategies for shutting down a nuclear power plant in the event of an emergency.

At the same time, attempts to reduce or eliminate intrinsic risk often merely displace the risk. "You end up moving it around," Leveson says. For example, medical X rays prove highly beneficial in the diagnosis of a variety of ailments, yet the overuse of X rays increases the risk of developing certain types of cancer. The evident benefits have to be weighed against the potential risks. Similarly, automating a process may reduce the chance of a human operator doing the wrong thing, but such a change merely shifts trust from one set of human beings to another. In the glass cockpit, for instance, responsibility moves from the pilot to the engineers and software specialists who designed and built the avionics electronics. In fact, Leveson observes, the root cause of an accident is often not at the site of the accident itself.

Leveson likes repeating a joke that has circulated widely in the aircraft industry: "Future automated airplanes will have only a pilot and a dog in the cockpit. The pilot will be included to reassure the passengers while the computers fly the plane. The dog is there to bite the pilot in case he or she touches anything." To which Leveson adds, "The pilot will also be there so that there's someone around to blame in case of an accident."

Indeed, Leveson bristles when she hears of accidents attributed to human error. Such a designation can actually sidetrack analyses of the causes of accidents, and the careless and arbitrary use of this phrase generally provides little insight into the multiple factors involved in an accident. In their paper on the Therac-25 defects, Leveson and Turner note, "Accidents are often blamed on a single cause like human error. But virtually all factors involved in accidents can be labeled human error, except perhaps for hardware wear-out failures. Even such hardware failures could be attributed to human error (for example, the designer's failure to provide adequate redundancy or the failure of operational personnel to properly maintain or replace parts). Concluding that an accident was the result of human error is not very helpful or meaningful."

Of course, some technologies do reduce risk, but people often end up cutting back the margin of safety so there is no real gain overall. One prime example is the development of improved collision-avoidance equipment, which warns pilots of close encounters with other aircraft. The widespread introduction of this technology has encouraged control towers to reduce the distance allowed between airplanes approaching or departing from airports trying to cope with traffic congestion in the sky. Leveson herself has served as a consultant in the massive software development effort required to make collision-avoidance systems work properly and safely.

Ironically, the addition of safety features can cause new problems, particularly as users of the systems become complacent. For example, overreliance on their instruments often makes pilots less alert to the possibility of midair collisions and other potential hazards. Similarly, NASA's first-class safety program, instituted after the deadly fire in 1967 that killed three astronauts in the *Apollo 1* capsule before launch, was so successful and later space shuttle flights so nearly routine that flight controllers and other personnel began making safety concessions. Rockets and spacecraft were even launched under conditions that were known to increase the risk of failure. One tragic result was the explosion of the space shuttle *Challenger* on January 28, 1986.

In fact, the magnitude and severity of accidents appears correlated with the prevalence of the belief that those particular accidents just can't occur. When the ocean liner *Titanic* went down in the north Atlantic in 1912 after striking an iceberg, the disaster caused more deaths than it might have because the engineers who designed the liner really believed that they were building an unsinkable ship. As a consequence, they supplied too few lifeboats for the number of passengers on board, figuring that in the event of the worst possible collision, the ship would surely stay afloat long enough to allow for onboard rescue by other ships.

It's also probable that the *Titanic*'s engineers focused on the wrong type of hazard. They built the ship with separate, watertight compartments so that the vessel would remain afloat even if several of the compartments were breached and flooded. But in 1993, a new analysis by maritime experts suggested that the *Titanic* had actually fractured, then broken in half, because of structural weaknesses in the ship's steel plates, which were fabricated from a brittle, low-grade steel. Chilled to freezing temperatures and stressed, these steel plates were primed to fail catastrophically—fracturing like glass rather than bending or stretching. Thus, even a minor collision with an iceberg could have caused major cracks to develop in the *Titanic*'s hull, with fatal consequences.

**NASA'S SPACE SHUTTLE** is one of the most complicated engineering projects yet attempted, and software plays a crucial role in this remarkable vehicle's operation. Looking like an ungainly glider, the space shuttle cannot fly during ascent or descent without a human pilot or an automatic control system making second-by-second adjustments to keep the craft on track. Taking into account a wide range of sensor data about the vehicle's physical state and its surroundings, shuttle flight software controls most aspects of ascent and descent, along with in-orbit operations.

NASA engineers are justifiably proud of the fact that the space shuttle software has never been responsible for a life-threatening failure aboard any space shuttle in NASA's fleet. The software's slow, methodical development and extensive, rigorous testing have prevented serious glitches.

But problems can arise. Glitches like the one described in Chapter 1 serve as a reminder that even heavily scrutinized, painstakingly developed software—all done at great expense—can still fail. And they underline the importance of maintaining a capacity for independently verifying and validating software.

This issue had come up in 1991, when NASA was contemplating as a budget-saving measure the elimination of the $3.2 million spent annually (out of a budget of $100 million per year on software development) on these independent software checks. That's not an unusual move. The testing of software is often one of the first casualties of any budget crunch—whether in the public or private sector.

However, before going ahead with its cutback, NASA decided to ask the advice of the National Research Council's Aeronautics and Space Engineering Board, which convened a panel of experts to evaluate the situation. Leveson chaired the committee, which issued an interim report in June 1992 strongly urging NASA to keep its verification and validation program intact. The report noted that this effort had in the past identified fifteen errors potentially serious enough to threaten the lives of a shuttle crew.

The panel's final report, *An Assessment of Space Shuttle Flight Software Development Processes,* was released on June 28, 1993. In essence, it commended NASA on its excellent work overall but argued there was room for improvement. In particular, NASA had not adopted the strict safety and process methods appropriate for such a large, complex, high-profile undertaking, in which the consequences of a serious mission failure would extend to loss of life, capital investment, and national prestige.

The panel members themselves found great difficulties in understanding what NASA had done and was doing. The reason was that much of what happened during the software-development process proved undocumented. There was a lack of adequately detailed, written descriptions of actions and decisions taken by the people involved. Instead, the panel members encountered a strong tradition of passing this "lore" orally from person to person, which put the whole process in jeopardy whenever key members of the development teams retired or left the agency.

Leveson titled a chapter of the report "The Silent Safety Program Revisited" to point out a link between the *Challenger* tragedy several years ago and shuttle software development practices today. In the official report on the *Challenger* accident, a chapter called "The Silent Safety Program" warned of a once-excellent NASA safety program that had become effectively "silent" over the years, perhaps due to complacency in the absence of accidents and a belief that less emphasis on safety was needed during operational use of the shuttle than during its original design. Leveson draws parallels between the practices that preceded the *Challenger* accident and those in the shuttle software development process today.

Whether anything comes of the panel's twenty-two recommendations depends on the priorities within NASA. Given the agency's numerous interests, including its battle to keep its space station project aloft, it seems unlikely that shuttle software improvements will rank high on the list. In the past, NASA has tended to ignore software improvement recommendations, partly out of stubbornness, partly out of lack of resources.

Leveson is a little more hopeful this time. "We tried to be very positive," she remarked soon after the report was issued. "And I think because of that, some of the people within NASA are very positive about the recommendations and taking this very seriously."

At the same time, the shuttle software—consisting of about three to four hundred thousand lines of code on board the spacecraft—is actually quite small compared to the software required for NASA's proposed space station. This program may run from three to four million lines. "You just don't get away with this entrepreneurial, everyone-do-their-own-part-in-their-own-way [approach] when you have such a complex program," Leveson says. "Having a good [software development] process is just crucial in whether you're successful or not."

One of the ironies in this field is that efforts to increase safety often make the entire system more complex, and that in itself can lead to accidents—especially when the system is so large that neither developers nor users can understand it well enough in its entirety. Moreover, any safety device, whether a fuse or a set of instructions in a computer program, can fail too.

Leveson frequently hears from companies that have built complex software without considering safety and then want her to prove that their systems or products are safe. "There is no way to pass a magic wand over software and declare it safe," she maintains. "If safety has not been designed in from the beginning, then any post facto procedure is a waste of time and money."

Leveson likes to point out that achieving improved safety can sometimes lead to a better product for reasons having nothing to do with safety. For example, the deaths of children trapped in refrigerators forced manufacturers to switch from a mechanical door latch to a magnetic fastener scheme as a safety measure. But the manufacturers soon found that the new device was also cheaper and more effective. In the end, the safer design proved to be the better design.

Expect things to go wrong, Leveson concludes, then do what you can to make sure that if something fails, it won't be catastrophic. That doesn't mean getting all errors out of software; it's impossible to do that except in the most trivial cases. There are always circumstances under which software will fail.

But perfection isn't necessary to prevent accidents. "This has been long recognized by hardware engineers who do not have a theoretical possibility of achieving failure-free operation," she says. "When software engineers eventually realize that software design perfection in complex systems may be just as elusive, they may also find that the alternative is to prepare for failures and to design to minimize their consequences."

Leveson is both a technological optimist and pessimist. "We can do better," she insists, adding, "I don't know if we can do it well enough."

Self-restraint is a quality notably missing from our headlong plunge into software for managing the world's business. "Where it is not possible to design the systems to be fail-safe and to protect against hazards, then the builders and users of these systems must be willing to accept a high level of risk or abandon the use of computers to control safety-critical functions," Leveson says. "It is difficult to accept that some systems cannot and should not be built, but it is also wrong to try to fool ourselves or others into thinking that we can achieve and guarantee ultrahigh reliability in software when all evidence is to the contrary."

How safe is safe enough? Ultimately, this is neither a scientific nor a mathematical question. It's a question of values—a political, cultural, economic, and moral matter.

# CHAPTER 3

# Power Failure

SITUATED BARELY AN HOUR'S DRIVE east of Toronto, the huge, boxy structures of the Darlington Nuclear Generating Station huddle on the shore of Lake Ontario. Completed in 1992, this modern power plant efficiently goes about its business of supplying enough electricity to serve a city of two million people. From its airy, sun-brightened visitor's center and neatly landscaped grounds recently cleared of construction debris to the freshly painted corridors and the control room's soothing electronic hum, the entire enterprise exudes an aura of quiet competence. It's the newest—and possibly the last—nuclear generating station to be constructed and operated by Ontario Hydro, the province's utility.

This plant is also the first Canadian nuclear station to employ computers to operate the two emergency shutdown systems that independently safeguard each of its four reactors. In each shutdown system, a computer program governs an array of electrically operated switches and relays, responding to sensors that monitor various parameters critical to a reactor's safe operation.

Ontario Hydro's decision to opt for software control carried unanticipated costs. To satisfy regulators that the shutdown programs would function as advertised, utility engineers had to go through a frustrating, tedious, and costly checking process to prove the system's trustworthiness. This task added more than a year to the time that it took the engineers to design and write the computer programs in the first place.

Still, the ten thousand or so lines of instructions, or code, required for each shutdown system pale in comparison to the hundred thousand lines of a typical word-processing program or the millions of lines needed to operate a long-distance telephone network or the space shuttle. The Darlington experience clearly demonstrated how even a relatively short, simple computer program can prove difficult to check. It also highlighted how strongly experts can disagree over what approach to take.

According to Canadian practice, an emergency shutdown system must operate independently of the rest of a nuclear plant. The emergency system receives data directly from sensors dedicated to this specific function, and it automatically overrides the regular control system to shut down the plant if any problem is signaled by these sensors. In older plants, such an emergency system relies on an electromechanical arrangement that compares the signal arriving from a sensor with a predetermined value. A signal larger than this value would automatically trigger electromagnetically controlled relays and switches that shut the reactor down. There are no computers in the loop.

When Ontario Hydro started designing the Darlington plant in the early 1980s, its staff was already considering the use of computer screens instead of dials and gauges as a superior means of presenting information, including sensor data, to the plant operators. With computers already included in the design, it didn't take a great leap to suggest that software could also control the shutdown systems. In fact, the designers thought it would be safer and cheaper in the long run to use a computer-based scheme, and it would allow greatly increased capabilities.

For example, making alterations to fine-tune the system or add new functions required merely changing a few lines of a program rather than modifying or installing bulky, cantankerous pieces of mechanical equipment. Computers could also do more checking, taking into account such factors as sensor failures. In essence, a computer-operated shutdown system appeared to offer greater economy, flexibility, reliability, and safety than one under mechanical control.

The contract for designing the two shutdown systems went to Atomic Energy of Canada Limited (AECL), a government-owned company that had pioneered the heavy-water type of nuclear reactor used in Canada. It was a company that took considerable pride in its technical expertise and its safety record, although the deaths and injuries attributed to malfunctions of its Therac-25 radiation therapy machine (described in Chapter 2) produced by one of its divisions that was later sold, have certainly dimmed that reputation.

AECL engineers developed the overall scheme for the two shutdown systems, and two separate teams wrote the necessary software. Originally, the design engineers had intended simply to duplicate in software the duties of the old, hardware system of switches, relays, and meters. But it proved impossible to resist the temptation to incorporate additional features and new monitoring functions that the use of digital computers allowed. The thinking was that such enhancements would make the system work even more effectively and safely. Thus, although these shutdown systems have a simple task, the computer-based versions that emerged after two years of design, coding, and testing turned out significantly more complex than the straightforward, easily inspected mechanical controls they replaced. Complicated pathways and shared data took the place of individual, obviously connected devices.

When the software was ready, Ontario Hydro had to obtain the approval of the Atomic Energy Control Board (AECB), the federal government body that regulates and licenses Canadian nuclear power plants. After examining the submitted material, AECB officials quickly realized that no line-by-line reading of the two programs would suffice to guarantee that they performed correctly. The programs were complex enough to warrant the use of outside experts to evaluate the software. To dig deeper, the board turned to David L. Parnas, a leading software engineer then at the Telecommunications Research Institute of Ontario, located in Kingston.

Such a gadfly role was not an unusual one for Parnas. Blunt, candid, outspoken, abrasive, and scrupulously honest, he is a well-known figure in the software engineering community. "People only call me in when they want me to tell the truth," he once commented. "So I'll tell them the truth, whatever it turns out to be."

PARNAS CAME TO THE AECB TASK with more than twenty-five years of experience in the concerns of software engineering. A native New Yorker and graduate of the Bronx High School of Science, he had started in physics at the Carnegie Institute of Technology (now Carnegie Mellon University) in Pittsburgh in the late 1950s. But it wasn't long before he became disillusioned with physics. Instead, Parnas turned to electrical engineering and obtained his undergraduate degree in 1961. Fascinated by computers, he progressed quickly. As a graduate student, he taught himself the basic principles of computer design from a leading textbook, then presented Carnegie's first course on the subject. Parnas also chose his own thesis topic and developed it without relying on an advisor, receiving his doctorate in 1965.

Gradually, his interests shifted from hardware to software. He became an activist of sorts for the methodical and structured development of computer programs. He acquired a strong appreciation of the value of breaking down a large job in just the right way to make individual programming tasks easier.

In the early 1970s, Parnas pioneered a style of programming that encapsulated computer programs into modules interconnected only in limited, carefully delineated ways. Using such a scheme, programmers could hide information inside a module without affecting how that module interacts with the rest of the program. Thus, a large program could be broken down not into successive steps, each one dependent on the one before, but rather into modules designed so that the details of their functioning are hidden from other modules.

This represented an important innovation, because software—unlike physical objects such as buildings or airplanes, which can be divided naturally into subunits with simple connections—has no obvious physical structure to guide its subdivision into logical units. Software is also exempt from physical constraints, which unfortunately encourages complicated interfaces between any modules built into programs.

Parnas's notion of information hiding has proved especially important during the design phase of software development. By restricting crucial design decisions to individual modules, a designer can build flexibility into the system. If a particular decision is later changed, then only one module has to be altered. The trick is to choose the right set of modules.

Consider the U.S. Navy's original computer-based system for processing messages, which acted like a seagoing telegraph office. Roughly speaking, the existing software could be divided into separate units for composing messages, storing and retrieving the messages from disks, and transmitting and receiving them. Such a design seemed sensible until the Navy decided to change the format of its messages to allow for longer addresses. It was like changing from five-digit to nine-digit ZIP codes.

Unfortunately, the system passed messages from module to module in a format with a fixed number of characters apportioned to each section of a message. A change in message format required changing every module in the system. A more sensible, flexible approach would have been to create one module for the message format, with specifications on how every other module interacts with it. It would then have been possible to alter the message format simply by reprogramming this one module without disturbing any of the others.

In essence, modular programming is an iterative process. The overall system is composed of a manageable number of modules, each one documented and hiding certain types of information. These modules are in turn broken down into sets of submodules that hide smaller secrets, and so on, until the final modules are so small that each one can be written by a single programmer.

As Parnas put it, "Constructing big programs without ever writing a big program is the key to constructing correct big programs." Such a strategy has become one of the crucial elements required for making software reliable.

While working as a consultant to the Naval Research Laboratory, Parnas got a chance to put his ideas into practice, and he ended up leading a project to redesign the onboard flight controls for the A-7E attack jet. It was a unique opportunity—the redesign of a major software system that already worked in an airplane about to be retired just to demonstrate that the new software was at least as effective as the old. Parnas noted at the time, "For the first time, we'll be able to compare a system built the right way with a system that performs the same task but is built the 'wrong' way." It was a luxury that only the Department of Defense could afford.

The project started in 1977 and was supposed to last about three years. Instead, it took eight years, partly because of budget constraints and partly because Parnas and his group had no ready-to-use model to guide them. They had to invent their own methodology as they went along.

When the project ended, Parnas could point with some pride that among the project's results was a complete description of the software requirements for the A-7E avionics system. The document was readable enough that it could be reviewed by pilots, which contributed considerably to its unusual accuracy, yet precise and detailed enough that it could be used by programmers as their only source of information about the requirements that their programs had to meet. It was a striking victory in the software trenches for Parnas and his team.

**PARNAS ACHIEVED** a certain notoriety in the mid-1980s when he became entangled in the nationwide debate concerning the feasibility of the Strategic Defense Initiative (SDI), or "Star Wars." On March 23, 1983, when President Ronald Reagan first announced his intention to proceed with SDI, he presented a compelling vision of an impenetrable shield of space-based weaponry that would shoot down incoming nuclear ballistic missiles before they could reach the United States. Indeed, the system was to have the capability of intercepting enemy missiles within ninety seconds of launch and destroying them before they could travel an appreciable distance toward their targets. It was easy to imagine that the resulting skirmish in the skies would be over in thirty deadly minutes or fewer.

Human reaction times, however, are too slow to manage such a battle. Only computers can process raw data from satellite-based sensors quickly enough to identify suspicious launches, pinpoint the sources of attack, compute the trajectories of enemy missiles, distinguish between warheads and decoys, assign targets to individual weapons, aim and fire the weapons themselves, and track the results to determine if additional action is necessary.

With so much to do, the SDI battle management system had to be far more complex than any body of software yet devised. As many as a hundred million lines of computer code, written by hundreds of individual programmers for a vast network of high-performance computers, would serve as its brain. It would have to operate successfully with practically no human intervention.

Even SDI officials freely acknowledged that developing the necessary software presented a formidable challenge. One report noted, "Specifying, generating, testing, and maintaining the software for a battle management system will be a task that far exceeds in complexity and difficulty any that has yet been accomplished in the production of civil or military software systems." It pointed out there was no complete, technical solution to the problem of testing the entire system under realistic conditions, short of a nuclear war. It admitted there was no technical way to guarantee absolutely that the system was safe and secure during times of peace, yet could survive wartime conditions.

But the report didn't see these concerns and potential shortcomings as insurmountable obstacles or reasons for questioning the wisdom of proceeding with SDI. Instead, these were issues to be addressed and problems to be solved. SDI proponents argued that the whole system was simply a bigger version of what had already been done, and human ingenuity would take care of the rest.

Early in 1985, the SDI organization invited Parnas to become a member of its panel on computing in the support of battle management. By that time, SDI was already a volatile, contentious issue, having prompted a chorus of skeptical responses. The panel's function was to provide advice to the Pentagon on what manner of computer-related research would be required for building the software at the heart of SDI.

Parnas was the one panel member closest to being a specialist in military software. He assiduously hunted down every piece of information that he could find on SDI. He couldn't help but bring a strong dose of skepticism to the battle management panel's deliberations. Initially, he had found the idea of a missile shield attractive, but he kept running into issues that pointed toward an excessively low probability of success, especially in creating the software necessary for the project. He couldn't understand how the other panel members could embrace the concept so fervently.

"People familiar with both software engineering and older engineering disciplines observe that the state of the art in software is significantly behind that in other areas of engineering," he later noted. "When most engineering products have been completed, tested, and sold, it is reasonable to expect that the product design is correct and that the product will work reliably. With software products, it is usual to find that the software has major 'bugs' and does not work reliably. No experienced software developer expects a product to work well the first time that it is used. Problems may persist for years and sometimes worsen as the software is 'improved.' "

Parnas quickly noticed that neither the SDI organization nor the public fully appreciated the problems inherent in computer systems. Most of the debate over SDI focused on strategy or hardware. The failings of software were rarely discussed. "I came to the conclusion that the public must be reminded of these facts while the debate about SDI was raging," Parnas recalls.

When Parnas dramatically quit the SDI computing panel on June 28, 1985, he issued a scathing letter of resignation, accompanied by eight short essays that called into question the entire SDI software effort. "My conclusions are not based on political or policy judgments," Parnas declared. "Unlike many other academic critics of the SDI effort, I have not, in the past, objected to defense efforts or defense-sponsored research. I have been deeply involved in such research and have consulted extensively on defense projects. . . . My conclusions are based on characteristics peculiar to this particular effort, not objections to weapons development in general."

Drawing on his extensive, firsthand knowledge of military systems, Parnas could readily point out where SDI software would inevitably fail. For example, even minor modifications of software for military aircraft require extensive ground testing and then flight testing under conditions closely approximating those in a realistic situation. Even then, bugs can and do show up in battle. To test even a part of strategic defense software under realistic conditions would require no less than aerial nuclear explosions, obviously an impossibility.

Similarly, it's commonplace for systems to undergo software modifications after they are deployed. In fact, it's not unusual to have programmers transported to field units to correct problems. "It is only through such modifications that software becomes reliable," Parnas maintains. There would be no such opportunities in a war that lasts only thirty minutes.

Experience with available early warning systems adds weight to these concerns. On June 3, 1980, at 1:26 A.M., the displays at the Strategic Air Command (SAC) post at Offutt Air Force Base near Omaha, Nebraska, suddenly indicated that two submarine-launched ballistic missiles were heading toward the United States. Eighteen seconds later, the early warning system signaled that even more missiles were en route. Then, strangely, it announced that no missiles were coming. Shortly thereafter, new information appeared showing that Soviet intercontinental ballistic missiles had been launched. Meanwhile, displays at the National Military Command Center (NMCC) in the Pentagon started to register the launch of enemy missiles from submarines.

These signals precipitated substantial preparations for retaliation, including the readying of bomber crews and land-based missiles. But while these measures were being put into effect, the top duty officers at SAC, NMCC, and the North American Aerospace Defense Command Center (NORAD) held a "threat assessment conference." It turned out they had enough additional information and time available to decide that all this screen activity was a false alarm. A similarly troubling incident occurred three days later, again initiating an alert.

Eventually, the trouble was traced to a faulty integrated-circuit chip in a computer used for communications to report analyses of NORAD sensor data to SAC, NMCC, and other centers. Normally, military personnel tested these communications links by sending fake attack messages, with a zero filled in for the number of missiles detected. When the chip failed, the system started filling in the space for the number of missiles with random digits. Unfortunately, the system designers had failed to incorporate any sort of scheme for detecting and correcting such numerical errors.

False alerts have also occurred for a variety of other reasons. For example, natural events such as moonrise or the flight of a flock of birds have confounded motion sensors. In another instance, an operator mounted a test tape on a tape drive that was mistakenly connected to the operational missile alert system, which caused the system to react as if a mere simulation were a real attack. Fortunately, in no known case did the United States come close to launching its missiles.

"Most importantly, human judgment played an essential role in the procedures followed in the event of an alert, and these procedures provided enough time for the people involved to notice that a computer system was operating incorrectly," computer scientist Alan Borning of the University of Washington commented in a 1987 article on computer system reliability and nuclear war.

But Parnas did more than outline why he believed the software portions of SDI couldn't be assembled to create a trustworthy system. He mounted a powerful attack on the status of software research, explaining point by point why each of the leading directions in contemporary computer science research offered no solace for SDI supporters.

Parnas contended, for example, that despite new techniques for improving the speed and accuracy with which software is produced, programming would remain largely a trial-and-error craft for a long time to come. In other words, despite the use of new programming languages that describe computer routines in broad terms, an emphasis on breaking up programs into smaller units that individual programmers can work on and then integrate, and introducing computers to help programmers keep track of their work, no sufficiently dramatic improvement in quality and reliability appears within reach.

Automatic programming systems, in which computers write computer programs, showed too little promise to be of much help for such a gigantic task as writing the SDI software, Parnas argued. Systems that accepted instructions and requirements at one end and produced code at the other end still relied on a programmer's skill in working out the problem first. And the resulting programs tended to be long-winded.

Parnas also faulted reliance on expert systems to guide programmers through programming intricacies. Expert systems are computer programs that incorporate a web of rules to solve a problem in much the same way that humans apparently solve it. "I find the approaches taken . . . to be dangerous and much of the work misleading," Parnas stated. "The rules that one obtains by studying people turn out to be inconsistent, incomplete, and inaccurate. Heuristic programs are developed by a trial-and-error process in which a new rule is added whenever one finds a case that is not handled by the old rules. This approach usually yields a program whose behavior is poorly understood and hard to predict."

Researchers in the field of artificial intelligence accept such an evolutionary approach to programming as normal and proper. Not Parnas. "I trust such programs even less than I trust unstructured conventional programs," he wrote. "One never knows when the program will fail."

To Parnas, there was no way in which anyone could guarantee that the SDI software could ever be trusted. "I am not a modest man," he declared in *New Scientist* on November 21, 1985. "I believe that I have as sound and as broad an understanding of the problems of software engineering as anyone that I know. If you gave me the job of building the system, and all the resources that I wanted, I could not do it. I don't expect the next twenty years of research will change that."

Though visible and colorful, Parnas didn't have the SDI software debate all to himself. Other computer experts weighed in on both sides. The panel from which Parnas resigned—which came to be called the Eastport group—issued its own report stressing that SDI software problems were surmountable. But the report came with a crucial change in premise, saying the system could be counted successful if it merely made any possible attack *unlikely* to achieve its goal. In other words, the SDI shield wouldn't need to be perfectly leakproof to act as a deterrent, and the software wouldn't have to be completely error free.

Led by computer scientist Danny Cohen of the University of Southern California, the Eastport group agreed that no massive, monolithic computer program organized like a military chain of command could possibly succeed. In place of an octopus with one brain and thousands of tentacles, the panel argued in favor of an architecture of numerous, loosely coordinated units that were largely independent of one another. There would be no centralized and detailed control over individual sensors and weapons. Control would rest largely with battle groups, each of which would act for the most part autonomously.

Cohen's group contended that the software for such a distributed system would be simpler and easier to write, test, and modify than software for a centralized arrangement. Also, the resulting computer programs would presumably be more resistant to sabotage, sensor or weapon malfunctions, battlefield uncertainties, and other potential failings. But this approach countered the understanding prevalent in software engineering that programming concurrent activities is more difficult than programming sequential ones.

Thus, even the adoption of a distributed style of computer architecture presented a considerable technical challenge. Just as every soldier on a battlefield has a different, and probably rather confused, impression of what is going on, an individual sensor-weapon unit might very well have too little information on the overall picture to judge the appropriateness of its actions. Moreover, how would a computer decide what to do when it receives contradictory messages? How would these machines cope with unreliable communications? How would they recognize casualties or malfunctions in other parts of the system and work around them?

All these considerations would have to be programmed somewhere into the system.

The Eastport group firmly believed that these difficulties, though considerable, were tractable, given sufficient effort. Parnas didn't. He fired back from his post at the University of Victoria in British Columbia (where he had arrived in 1982), "[A] careful reading of the Eastport report reveals no basis for its more optimistic conclusions." Parnas asserted that most of the objections that he had raised against the overall SDI software scheme applied just as well to the software required for controlling individual, independent battle stations. He also made it clear that he couldn't accept the assumption implicit in the Eastport report that local, seemingly minor errors among networked computers could not cause widespread failures. There were too many examples already of just such faults in large, complex computer systems, he maintained.

Finally, Parnas reiterated his key point: He had never argued against SDI with the expectation that its software would have to be completely error free. "I have consistently used the word 'trustworthy,' explaining that what is required is a system with known, that is, predictable, effectiveness—one that we know, with great confidence, is free of catastrophic flaws," he remarked. "That is far from perfection."

He added, "My conclusion about SDI was not that it would not be leakproof but that it would never be trusted. We would never be able to believe, with any confidence, that we had succeeded in building an effective shield." It was up to computer professionals to tell the public honestly what it could realistically expect to get out of the nation's immense investment in SDI software.

But the sniping continued over the next few years, sporadically flaring into bitter confrontations and debates among professionals and between professionals and policy makers. At the same time, a sizable number of computer scientists were ready to offer their expertise to solve the perceived problems.

Parnas now likes to play down his role in the SDI debate, insisting that his involvement constituted an unimportant phase of his life and career, though it was something he had to do. He maintains that this effort represented a minor detour from his mainstream interests, though ironically it was the military that served as the primary funder of the research he wanted to do. But the controversy gave him an unmistakable public profile, and it brought out his sense of professional ethics and his independence of mind.

BY 1987, Parnas had taken a new position at the Telecommunications Research Institute of Ontario. When the Atomic Energy Control Board came to him asking for help with its evaluation of the Darlington emergency shutdown system software, his first reaction was to say no. He was tired of controversy, he told the AECB.

But in pondering the situation, Parnas got out his old SDI papers, which listed the reasons why the SDI software couldn't ever be trusted, and he decided that none of these criteria applied to the Darlington case. Indeed, if he really had faith in the techniques he had pioneered for checking software, the AECB call for help provided just the kind of platform he could use to prove that his methods actually worked on a real-world example.

Nonetheless, Parnas's first glimpse of the Darlington programs wasn't encouraging. "When I looked at the code, it became clear that it was impossible for me to say whether it was right or wrong," Parnas later recalled. "All I could say was that the documentation was too vague." In other words, from his point of view, the Ontario Hydro and AECL engineers who had written the two shutdown programs hadn't done an adequate job of describing what each part of the program was supposed to do. It was all expressed in English-language sentences that left room for ambiguities and misunderstandings.

"I didn't know a thing about nuclear power plants, and I didn't have the requirements to work against," Parnas says. He asked for better documentation, insisting that otherwise he couldn't evaluate the software.

Ontario Hydro and AECL engineers objected strenuously to having to spend extra time on what they considered an unproductive exercise just to please a lone software engineer out of academia and his graduate student assistant. They firmly believed that they had a viable system and that although the software was hard to review, it was basically correct. They couldn't understand why the AECB was taking Parnas seriously while ignoring the assurances of the dozens of experienced engineers who had worked on the project.

AECB, of course, was used to this kind of resistance. It occurred every time the regulatory body insisted on a particular course of action. This time, its concerns were bolstered when Parnas, in the course of trying to persuade the engineers that the documentation was too vague, stumbled upon an unsuspected ambiguity in the specifications that could potentially jeopardize safety.

In later descriptions of the problem, Parnas was not allowed to give any details of the existing shutdown systems, so he made up an example that illustrated the type of ambiguity he had found in the code: Suppose the nuclear engineers had included a parameter based on the water level in the reactor for triggering a shutdown. Their specification might state: "Shut off the pumps if the water level remains above one hundred meters for more than four seconds."

This sentence appears straightforward, but what if the water level varies during the four-second period? Parnas came up with three different interpretations of this statement, based on different ways of finding the average water level. A programmer could choose only one of the three. Which one was appropriate?

From the software, Parnas discovered that the nuclear engineers had interpreted this specification in a *fourth* way, which differed from the three choices he had suggested. Expressed more precisely, what they had done in both the main and backup shutdown systems corresponded to the statement: "Shut off the pumps if the *minimum* water level over the past four seconds was above one hundred meters."

To Parnas—and to AECB officials—this fourth interpretation contained a flaw. With sizable, rapid waves in a tank, the water level could be dangerously high but not trigger a control system built according to this specification. To the nuclear engineers, however, this was such an unlikely possibility that it hadn't been taken into account.

The ambiguity revealed in the specifications for the shutdown software, involving not water level as in the example but some other crucial parameter, was enough to persuade the AECB to insist on a thorough software review, starting with improved documentation of the requirements. The engineers still objected. In their view, Parnas was an outsider looking at a process from the point of view of a programmer who knows nothing about a system to start with and must write code to meet requirements as the only way to get the software right.

That clearly didn't apply to the Darlington situation, they argued. Even though specifications weren't written down in a mathematically precise way, an experienced designer of shutdown systems would know what they meant and could convert them correctly into the lines of a computer program.

Parnas argues that the wrong thing was still done at Darlington—even by people who did know the field. Moreover, only one of the two systems was written by nuclear engineers. Because none of the engineers knew the particular programming language in which the second shutdown system was to be written, Ontario Hydro and AECL had to bring in outsiders to write the software for it.

"We had a difference of approach here, which [Parnas] didn't appreciate and we failed to communicate," says Glenn Archinoff, one of the Ontario Hydro engineers involved in the development and testing of the Darlington software. "That difference in understanding of how things should be done was the start of a lot of problems."

But G. J. K. Asmis, the AECB engineer who oversaw the software-assessment program, recognized that experienced nuclear engineers didn't necessarily know how to write high-quality code. "The programs were put together by engineers who were not software engineers. They used their own personal ideas," he noted. "Their ideas of modules, for example, were not the modern ideas of information hiding, and so on."

An immediate and contentious issue concerned the adequacy of the testing that had been performed. Ontario Hydro and AECL had already systematically subjected their software to a large number of carefully constructed tests designed to ensure that it would function properly under a variety of circumstances. But such planned tests cover only a fraction of the possible paths through the software, and they often miss subtleties.

Parnas insisted that Ontario Hydro also perform random testing. He suggested, for example, that the utility furnish randomly generated sensor data, with values distributed statistically to match the kind of inputs the system would experience, to see how the software would respond.

"One of the funny things about the field is that there are good arguments that say you're better off doing random testing than carefully controlled testing," Parnas says. "But that's not an intuitively obvious thing." In fact, when Parnas himself first encountered the idea more than a decade earlier, he had thought it was crazy. Only after careful review did he accept this approach as fundamentally sound.

In a 1990 paper presented at a conference on critical issues in computing, Parnas and Asmis noted, "[T]here is good reason to believe that those who design the planned tests will overlook the

same cases that the program designers overlooked. This is particularly true when programmers test their own programs. . . . Random testing is free of the prejudices that cause human testers to
overlook certain cases."

Ontario Hydro engineers ended up doing a significant amount
of random testing (though not as much as Parnas would have
preferred) to complement their original testing program. On
one of the few occasions when Parnas visited the plant, he was shown
the testing and told that no errors had yet been found. The
engineers then generated a test on the spot and took Parnas
up to the control room to see the results. There he learned
that clerks were checking the results manually. Parnas glanced
at the sheet of paper generated by the latest test and after a
moment, pointed out that something wasn't functioning
properly. It turned out that one reason no errors had yet been
picked up was that the clerks were far behind in checking the
numbers.

In the end, the tests turned up several minor problems. Corrections for these discrepancies, along with changes mandated by
revised requirements, went into the second version of the shutdown software. Most of the time spent on the new version, however, involved rewriting the documentation to better reflect the
software's functions.

At this stage, to help establish the safety of its shutdown systems, Ontario Hydro brought in a consultant, software safety expert Nancy Leveson from the University of California at Irvine.
Leveson's approach was to look not at the specifications—which
were bound to be incomplete, in her view—but at what the system was actually supposed to do. An emergency shutdown system,
for instance, is there to shut down the reactor, and the system fails
if it neglects to do just that.

Leveson recommended that Ontario Hydro perform a hazard
analysis. The idea behind such an analysis is to consider all the

ways in which a system can fail, and then to work backward through the hardware and software components to determine what factors could cause such failures. Designers are thereby able to incorporate the appropriate safeguards.

Leveson found the basic design of the resulting software simple, safe, and sound. In fact, the shutdown systems incorporated a particularly useful feature that had also been present in earlier electromechanical versions. It worked something like this. Suppose there were eight switches, and the system shuts down if any one of the switches is in the off position. Assume all the switches start in the on position. Then, if any one of them receives a signal indicating a problem, that switch will go off, and a shutdown will occur. But it's actually much safer to start with all the switches in the off position. Then, if all the switches receive signals indicating that all is well, the switches will turn on, and no shutdown will occur.

This strategy forestalls a scenario in which something bad happens but no signal reaches the appropriate switch. In the first case, the system continues to operate despite the problem. In the second case, the switch stays off, and the system shuts down. The nuclear engineers had adopted this fail-safe design to make sure that their systems would function safely.

To Parnas, however, the hazard analysis wasn't an adequate safety check. He believes that every piece of a safety-critical system is crucial and must be checked mathematically. "In our experience, software exhibits *weak-link* behavior; that is, failures in even the unimportant parts of the code can have unexpected repercussions elsewhere," Parnas and his collaborators argued in a 1990 paper on the evaluation of safety-critical software. Safety can't be separated from reliability and trustworthiness.

As a result of Ontario Hydro's hazard analysis, the personnel responsible for the review made forty-two changes, most of them involving little more than altering the order of program statements and including extra checks for detecting potentially hazardous situations. The third version of the Darlington software incorporated these changes and the results of additional random testing, along with modifications due to shortcomings pinpointed by comparing Ontario Hydro's process for developing software against a newly promulgated European standard for assuring quality in software engineering.

The AECB then called for a full, formal verification of the third version of the software, applying a method that Parnas and his colleagues had developed. In essence, this meant expressing each of the functions performed by a computer program in a mathematical form, then comparing this mathematical description against the software requirements, also expressed in a mathematical form.

This systematic inspection proved both exhausting and exhaustive. Four separate teams went to work. No one lasted more than six weeks. Fresh personnel regularly arrived to take over and continue the laborious checking.

The process turned up hundreds of discrepancies, but nearly all were harmless. In many cases, the computer programmers had inserted extra instructions, such as additional safety checks, which were not called for in the specifications. The teams also found a handful of coding errors, but none of the errors proved serious enough to delay or prevent an emergency reactor shutdown. In the end, although there were many changes, the computer programs remained essentially the same as they had started. "The engineers had put a lot of effort into trying to get it right, and basically they succeeded," Asmis says. Nonetheless, the changes improved the system's overall safety.

Three years after the start of its inspection, AECB finally gave its approval, and the Darlington plant received its license to start up its first reactor in February 1990. At the end of the lengthy, tedious process of reviewing the software, the regulatory body no longer had any reservations about the utility's ability to shut down the reactor in an emergency.

Looking back, Archinoff and other Ontario Hydro engineers

can actually point to some benefits that came out of their ordeal. In going through the laborious process, the engineers ended up with a more thorough understanding of the software. Unfortunately, the information resides in a highly mathematical form in more than thirty binders packed with tables and data from the formal evaluation. "That large amount of paper is a record of the mathematics of the problem," Archinoff says. But "it would be difficult for a third party to come in and know what to do with it."

Although the year-long mathematical verification proved beneficial to the engineers at Ontario Hydro, it was also extremely costly, labor-intensive, and time-consuming. Altogether, the effort required approximately thirty-five person-years of labor, mainly because it all had to be done by hand. The evaluation teams had no computer programs that could aid them in deriving the necessary tables and making the comparisons.

At the same time, many of the frustrations of the checking process could have been avoided if the software designers had written the programs with review in mind. "We encountered many problems simply because the designers and programmers had not expected to have their work verified by mathematical techniques," says Richard P. Taylor, one of the AECB reviewers.

Many of the engineers firmly believed that if they knew they would have to go through such a process again, using the same methods, they would have gone back to the old electromechanical systems instead. But these systems had their own faults and vulnerabilities. For example, they were more easily damaged and parts were harder to repair and replace than their software counterparts.

Engineers like Archinoff recognized the advantages of computer-based electronic control, and they didn't want to lose the benefits of using software in safety-critical systems. "Our challenge now is to develop techniques and tools that will enable us to do all of these more sophisticated, more formal processes but in a more cost-effective way," Archinoff says. "We're making progress on that."

Indeed, the software development effort didn't end with the startup of the power plant. When the AECB issued the license, it added this proviso: "The Board . . . has concluded that the existing shutdown system software, while acceptable for the present, is unsuitable in the longer term, and that it should be redesigned. Before redesign is undertaken, however, an appropriate standard must be defined."

One of the board's main concerns was what to do if the shutdown software were modified in some way. It had no confidence that changes could be made without the risk of introducing new errors. But like Ontario Hydro, the board didn't want to go through the same formidable exercise all over again, just because of some changes. It was much better to redesign the software, taking into consideration everything that had been learned so far.

It's traditional in Canadian nuclear licensing for the AECB to set the overall requirements and for the licensee to present to the regulator its plans for meeting those requirements. Hence, Ontario Hydro and AECL took on the task of developing methods for specifying, designing, and verifying safety-critical software, with a goal of making the software easier to modify and easier to verify and review. The AECB started working with Parnas and others to develop overall Canadian standards for safety-critical software in nuclear power plants, carefully monitoring international developments in this area to ensure that Canadian standards were on par with the rest of the world. Once these standards were in place, Ontario Hydro could go ahead with redesigning the software for its Darlington shutdown systems.

**THE DARLINGTON EXPERIENCE** with safety-critical software is not yet common in the nuclear industry, although software-safety problems are becoming evident. Early in 1991, for example, severe software problems forced Electricité de France to consider abandoning a complex, computer-driven "plant supervisor" developed for its new pressurized-water nuclear reactor. According to one French official, the system had to handle so many data concerning how the reactor was functioning that the computer program grew too complex for it to be modified easily when needed. Managers feared that the system would never pass inspection.

Software problems just as serious have emerged for the new Sizewell B nuclear power station in Suffolk, England. It's the first plant in Great Britain to rely heavily on computers for control and emergency shutdown, with both conventional and software-based protection systems. Far more elaborate than Darlington's shutdown software, this system uses several hundred microprocessors, commanded by computer programs totaling more than a hundred thousand lines of code.

But all has not gone well in building the system. By 1991, after eight years of development, several software engineers familiar with the project were ready to suggest that the software be scrapped because it had become unmanageable and that the whole system be built again.

An incident at the Sellafield reprocessing plant in Britain highlighted the potential danger. Designed to encase highly radioactive nuclear waste in glass blocks for transport and storage, this facility went into operation in February 1991. But in September of that year, a bug in the computer program controlling the plant caused radiation safety doors to be opened prematurely while highly radioactive material was still inside one chamber. Although no one was exposed to radiation and the plant was shut down immediately until the problem was fixed, the incident set off alarm bells at the Nuclear Installations Inspectorate (NII), Britain's nuclear watchdog.

The inspectorate had originally judged the plant's safety-critical software acceptable, partly because it consisted of only a limited amount of code. But the computer program was later amended, and it was this software "patch" that was apparently responsible for making the doors open too soon. The plant's operator, British Nuclear Fuels Limited, had believed that amending the software would have no impact on safety. Two years later, following the accidents, the company faced prosecution for making what was alleged an unauthorized software change to a safety mechanism on a shield door.

As for Sizewell B, software problems continued to plague the nuclear plant as it neared its completion date in 1994. To test the software responsible for shutting down the reactor in case of emergency, the plant operator had developed test software that would reproduce the wide variety of signals and messages the system might receive during an alert. However, in nearly half of the fifty thousand tests run, the shutdown system software did not give the expected responses, as listed in the specifications. The company blamed the test software for the failures, contending that the shutdown system itself was safe. That wasn't good enough for the Nuclear Installations Inspectorate, which insisted that the system be checked again.

In the United States, most existing nuclear power plants use computers only for functions unrelated to plant safety. However, officials at the Nuclear Regulatory Commission in Washington, D.C., believe that software control will inevitably creep into nuclear plant designs, and they are starting to prepare for the formidable task of software evaluation.

Beyond the nuclear power industry, software reliability considerations have a direct impact on chemical plants, oil refineries, and many factories. As described in previous chapters, they also play crucial roles in telephone networks, avionics software, and medical equipment.

"The nuclear industry is not my main interest," says Parnas, who in 1991 took up a position at the Communications Research Laboratory at McMaster University in Hamilton, Ontario. "I see it as only one aspect," he notes. "I have to ride on an A320 [airliner], and I'm not comfortable with what I've heard about the A320 [avionics software]. I have to use the telephone, and the network can fail."

His own life may now depend on a what he calls an "embedded piece of software." In June 1993, surgeons implanted a microprocessor-controlled pacemaker to regulate his heart. Parnas couldn't help pondering the disturbing notion that the necessary code had likely been written by people who had never taken a substantive software engineering course and who probably didn't use mathematics at all in thinking about software. Moreover, because the manual accompanying the device was poorly written and incomplete, Parnas found he had to advise the doctors on how to adjust the device.

Parnas ranks computer programs among the most complex products ever devised by humankind. They are also among the least trustworthy, he says. "These two facts are clearly related. Errors in software are not caused by a fundamental lack of knowledge on our part. In principle, we know all that there is to know about the effect of each instruction that is executed. Software errors are blunders caused by our inability to fully understand the intricacies of these complex products."

At the same time, most computer programs escape the kind of intense scrutiny applied to the Darlington shutdown system. The process is both costly and time-consuming, and the majority of people who write computer programs lack the expertise required to use the sophisticated methods necessary for producing reliable software. Furthermore, many problems arise not because of errors in programs but because of flaws in the specifications for what a given program should do.

Software testing and verification is itself the subject of considerable controversy. There's no general agreement on the best way to proceed and how effective various methods are. In the Ontario Hydro case, the Parnas approach generated a huge volume of paper filled with detailed tables purportedly demonstrating mathematically that the computer program's actions matched the system's specifications. But there was no guarantee that the information in these binders was all correct. Who, other than the people directly participating in the process, could slog through this mathematical morass to check the checkers?

Many times people don't know that they have run into the consequences of a software error—when a telephone call fails to get through, for example, or a traffic light behaves erratically, or an automobile suddenly accelerates when it should be coming smoothly to a stop. "You may see something unexpected happen, but you don't always think to question the software that may be involved," Parnas says. By sharing information, people become increasingly aware that problems with software and computer systems are not unique, unusual experiences.

"If we're on the brink of anything, it's on becoming aware of things going wrong," Parnas suggests. "One of the things I noticed in the 'Star Wars' debate was that everybody I talked to had experience with software errors, but they thought that their experience was unusual."

When Parnas was working part-time for the Navy in the late 1970s and early 1980s, he learned first-hand of this blindness to software defects via a series of accidents involving F-18 jet fighters, which would mysteriously end up crashing into Chesapeake Bay. It turned out that the cause was a software error, but for the first three accidents in the series, no one ever suspected a computer problem. Officials typically attributed the accidents to pilot error. It took a long time for them to figure out that the fighter's avionics software had forbidden the plane to take a certain action when demanded by the pilot because a programmer had once been told that you should never let the wing surface get into a particular position. In the test pilot's standard routine, there was one maneuver in which the plane got into such a peculiar spin that the wing surface had to get into the "forbidden" position for the aircraft to emerge from the spin.

"We can no longer treat software as if it were trivial and unimportant," Parnas says. But progress is slow. Writing computer programs still relies more on tradition and habit than on any scientifically based, rational approach. It is difficult to make accurate predictions of software reliability. It is not practical to measure the trustworthiness of software beyond a basic level. Standards are often incomplete, out-of-date, or just meaningless. Many programmers lack the qualifications and expertise to do the job required.

"So it's a pretty discouraging thing," Parnas concludes. "On the other hand, I stay in the field because I'm convinced that computers can be of real benefit to society. There are hundreds of things you can do with computers that you could never do without them."

In 1993 Parnas wrote, "Technology is the black magic of our time. Engineers are seen as wizards; their knowledge of arcane rituals and obscure terminology seems to endow them with an understanding not shared by the laity. The public, dazzled by the many visible achievements of modern technology, often regards engineers as magicians who can solve any problem, given the funds." The danger lies in taking all this power, convenience, and complexity for granted, and forgetting the fallibility of programmers, software engineers, and computer system designers.

# CHAPTER 4

# Experience Factory

**AS SMOOTH AND FEATURELESS** as glass, the delicately tinged, bluish green visage of Uranus has become a familiar image. So too has the deep blue face of Neptune with its Great Dark Spot, hazy bands, and traces of bright, wispy clouds. Prominently displayed in countless publications, pondered in dozens of scientific papers, and featured in computer games and posters, these images were captured by instruments aboard the robot spacecraft *Voyager 2* during an odyssey that took this spindly vehicle on a twelve-year trek across the solar system along a route of more than four billion miles. No Earth-based telescope could have produced such intimate, spectacular portraits of these mysterious, alien bodies.

Computers played crucial roles in this phenomenally successful venture. Indeed, the *Voyager* program kept numerous computer scientists and software engineers busy during the spacecraft's years in space. Computers supervised the August 20, 1977, launch of the Titan-Centaur rocket bearing the spacecraft. Innumerable trial calculations of possible trajectories allowed project managers to select a route that permitted *Voyager 2* to use the giant planets along its path as gravitational slingshots to change the vehicle's speed and direction to reach each successive destination on time and on target. All the while, *Voyager 2* remained tethered to Earth by a tenuous, infinitely flexible thread of electromagnetic radiation—radio waves—along which project engineers could transmit instructions to maneuver and control a vehicle that was otherwise beyond human reach. A computer-operated network of radio antennas on Earth picked up the extremely faint signals emanating from the spacecraft, and ground communications systems transmitted the digital information to data centers for processing into images.

*Voyager 2* itself carried aloft three pairs of computers, all powered by electricity generated by heat released from the decay of radioactive plutonium. A guidance computer and its backup controlled the spacecraft's orientation in space. A second pair processed the scientific and engineering data compiled by *Voyager*'s instruments before that information was broadcast back to Earth. A third set functioned as the overall coordinator of the spacecraft's actions and itinerary.

Instructions stored in these onboard computers kept the space probe on target and its cameras precisely trained on the dimly lit surfaces of Uranus and Neptune as it flew by at tens of thousands of miles an hour. These instructions could be brought up to date, changed, or replaced at any time with new commands radioed from Earth.

With *Voyager 2* on "automatic" for lengthy periods, its computers constantly monitored the spacecraft and the computers' own circuitry and software for malfunctions. Often, the computers could find errors and solve problems before human controllers at the Jet Propulsion Laboratory (JPL) in Pasadena, California, could become aware of them, let alone react. This capability, along with the flexibility that programming and reprogramming a computer allows, helped save the project from disasters that more than once threatened the mission. In several instances, the spacecraft's computers figured prominently in both the problems and the solutions.

Shortly after launch, ground controllers received a signal from *Voyager 2* that a boom was not fully extended from the craft. *Voyager* engineers checked the data carefully, ran tests, and finally decided there was no problem. The boom was actually in its correct position. The *Voyager* team reprogrammed the coordinating computer to ignore the spurious report from the boom sensor.

There were also major malfunctions. Nearly eight months into its mission, *Voyager* suffered the consequences of two bad electrical connections, which severely cut back the spacecraft's ability to communicate with Earth. The coordinating computer had been programmed to switch to a backup radio receiver if *Voyager* heard nothing from Earth over a one-week period. Such a silence could mean that the probe's primary receiver had failed. Then, after a twelve-hour wait, the computer would switch communications back to the primary radio receiver—just in case there had been nothing wrong with it in the first place.

In the first week of April 1978, *Voyager 2* received no messages from Earth. Distracted by problems with a companion spacecraft and perhaps lulled into complacency by how smoothly the mission had run to this point, the flight team at JPL simply forgot to call *Voyager 2* at any time during that fateful week. As instructed, the coordinating computer concluded that the primary receiver was faulty and switched to its backup. But the long-dormant backup receiver had a defective circuit that no one had detected. This circuit normally allowed the receiver to scan across a broad range of frequencies to zero in on a signal from Earth. After the circuit failure, the damaged receiver could detect only one frequency, and if a broadcast from Earth failed to hit this particular frequency, no message could get through.

To its operators on Earth, the spacecraft appeared to have turned itself off. It took project engineers hours to figure out what the problem was, and by then it was too late. Following its instructions, the coordinating computer aboard the spacecraft switched back to the primary receiver after the requisite twelve hours. This time, something went wrong in the primary unit—possibly a damaging electrical surge that caused a short circuit in its power supply. The receiver failed for good. The spacecraft was effectively deaf.

A week later, on cue, *Voyager 2* switched to its faulty but still-functioning backup. The backup receiver could detect radio signals, but the exact frequency at which it operated would wander as the spacecraft sailed through space and its temperature changed. Merely turning on several instruments at once or using one of the spacecraft's tiny thrusters for maneuvers would generate enough heat to send the receiving frequency scooting to some new value.

For the remainder of the mission, keeping in touch with *Voyager 2* became a major task and a delicate business. To reestablish contact with the spacecraft, the control team at JPL would try a number of different frequencies in the hope of hitting the right one—often waiting up to forty-eight hours after a heat-generating maneuver before making an attempt. Eventually, engineers worked out elaborate computer-based techniques for predicting the receiver's frequency at a given moment.

Even before *Voyager 2* reached Saturn in August 1981 and made its grand turn toward Uranus, project managers had gone to work improving the spacecraft's ability to transmit images of its targets. A single picture consisted of eight hundred lines of eight hundred pixels each, with a pixel represented by eight bits of data and showing up as a dot of a certain brightness on a monitor. Each picture required more than five million bits of information. But as the spacecraft traveled farther from Earth, its signals became weaker, and these bits had to be transmitted at lower rates. Thus, by the time it reached Uranus, *Voyager* would require ten minutes—an unacceptably long time—to transmit one image.

The solution was to compress the digital data into a smaller package. New software, radioed up to the spacecraft bit by bit, changed the imaging procedure so that only the first pixel in a line recorded the actual brightness level seen by the camera, while the remaining pixels in the line represented not the true brightness, but the differences in brightness between their levels and that of the initial pixel. The new strategy cut the amount of transmitted information required to reconstruct a single image on Earth by as much as 60 percent. To accomplish such a feat, one of the two data computers was devoted solely to imaging and compression, while its original backup handled more routine data-collection tasks.

With those and other improvements, *Voyager 2* became a more capable vehicle than it had been at the outset. In 1986, Ellis D. Miner, assistant project scientist for the mission, remarked, "For a spacecraft designed and built back in the days when hand-held calculators were first being marketed, the *Voyagers* [*Voyager 2* and its twin, *Voyager 1*] have been remarkably responsive to the science and engineering demands placed on them." Software had a lot to do with that success.

During the 1980s, ground-based systems were undergoing their own overhauls. New equipment and software went into the Deep Space Network, which consisted of three tracking stations positioned in California, Australia, and Spain. Antennas at each station picked up signals from *Voyager 2* and transmitted the compressed data to an imaging center, where computers performed the steps required to turn digits into pictures.

Of course, the changeover didn't go smoothly. There were serious bugs in the antenna and image processing software, which had to be corrected, even as *Voyager*'s January 1986 encounter with Uranus crept inexorably closer. At one point during the preparations, *Voyager* project manager Dick Laeser commented, "Now we're seeing funny little things in the data, and the trick is to find where the problems are in this huge new system, and then isolate and solve them. None of the problems yet are biggies, showstoppers. But we need to clean it up so we can trust the data."

When the time came to photograph Uranus, only one major imaging problem cropped up. On January 18, 1986, ground personnel started seeing peculiar black-and-white lines running through the images coming from *Voyager 2*. Step by step, engineers eliminated various facilities on Earth as the source of the error and proved that it had to be at the spacecraft. To locate the problem, they instructed the data compression computer to skip a picture and instead send down the complete contents of its memory.

In this maze of digits, the engineers found evidence that a memory location had become stuck. It no longer switched between one and zero but stayed at a fixed value, which was enough to disrupt the data compression process and put streaks in the final images. The engineers wrote a short sequence of instructions to detour the computer around the stuck bit, and the computer resumed sending compressed images to Earth.

The *Voyager 2* mission, despite potentially crippling problems, owed a great deal of its ultimate success to a remarkable coupling of human ingenuity with the malleability of software. Indeed, the same combination of ingenuity and flexibility is required for any kind of space endeavor, whether it involves a communication satellite in orbit around Earth or a sophisticated vehicle like the Jupiter-bound *Galileo* spacecraft. Such ventures would be impossible without computers and software for orbit calculations, vehicle design, navigation, ground control, communications, maneuvering, telemetry, and other functions associated with the vast, costly infrastructure of space exploration.

THE NATIONAL AERONAUTICS AND SPACE ADMINISTRATION'S Goddard Space Flight Center is the focus of many space-exploration activities. Created in 1959, this center for space-based research sprawls across land originally taken from the federal government's Beltsville Agricultural Research Center on the outskirts of Washington, D.C. It serves as the nerve center of NASA's extensive worldwide ground and space communications system and as home base for a wide variety of Earth-orbiting spacecraft. *Explorer VI*, the first satellite launched under Goddard supervision, provided

Earth with its first self-portrait from space. Subsequent satellites radioed back data on Earth's magnetic and gravitational fields, on radiation belts girdling the globe, on the solar wind of particles sweeping out from the sun, and on Earth's fragile environment. In one recent triumph, the *Cosmic Background Explorer* satellite provided the first details of a fledgling universe just starting to emerge out of a primordial electromagnetic fog.

Goddard is organized into directorates, the largest of which is devoted to mission operations and data systems. It consumes nearly a quarter of Goddard's $2.5 billion budget and produces about half of the software used at the center. It includes the flight dynamics division, where software engineer Frank McGarry manages the development of computer programs for controlling and maneuvering Earth-orbiting spacecraft. This software determines when a spacecraft should be launched and into which orbit it should be put. It tracks the satellite, keeping tabs on its orientation in space and its location at any given moment along its orbit. It calculates how long the spacecraft can remain aloft before crashing to Earth.

In the 1970s, like nearly every other developer of complex software, McGarry faced considerable difficulties in delivering a high-quality product on time and within budget. It was hard to predict with any degree of certainty when a project would be finished and how much it would actually cost. Moreover, the "completed" programs would inevitably contain bugs, which would have to be corrected, and the corrections would often lead to more bugs. Charged with developing about a half dozen new programs every year, McGarry wondered whether there was a demonstrably better way of producing the software.

There was no shortage of proposals. In the early 1970s, the desperate need for a more disciplined, orderly approach to writing computer programs had seen a movement toward structured programming. By the middle of the decade, that movement had spurred interest in pinning down and precisely defining the crucial elements of software development. There were new programming languages, new programming styles, and new management techniques. One research paper after another extolled the virtues of a particular procedure, style, technique, or set of rules for taming the software "monster." Any one of these methods would lead to the promised land of improved productivity and higher quality, their advocates insisted at seminars, workshops, conferences, and tutorials.

Still, programming itself remained very much a craft, subject to the vagaries of human thought and predilection. The notion of software engineering—the practical application of scientific knowledge in the design and construction of computer programs and the associated documentation required to develop, operate, and maintain them—was still new and largely unfamiliar to the programming community. Coping as best as they could, most programmers found themselves caught in a vicious spiral of write, test, debug, write, test, debug, and so on, never knowing quite when to stop.

Confronted by a confusing array of options for producing software, McGarry wanted proof that a particular approach or technique was really better than another before he would commit himself to changing procedures at Goddard. For McGarry, there was too much at stake to risk plunging blindly into something new and unproved.

In early 1976, McGarry went to Victor R. Basili, a computer science professor at the University of Maryland, to see if Basili and his colleagues would be interested in studying how software is produced in a real-world situation and in assessing which promising new techniques should be applied at Goddard. It was an offer and a challenge that Basili could not resist.

BASILI, A NATIVE NEW YORKER started out studying mathematics, and his initial encounter with computers was distinctly inauspicious. As an undergraduate at Fordham College, for a course on numerical analysis, he was required to write a short program on a strip of paper tape, which was fed into a Bendix G-15 computer. It was no fun punching holes in tape that readily ripped and jammed in the machine. "I thought computers were the stupidest things ever," he recalls. Nonetheless, while at Syracuse studying toward his master's degree, Basili ended up working in the computer center.

When Basili taught mathematics at Providence College in Rhode Island for four years starting in 1963, he became increasingly interested in exploring the mathematical basis of computing. Once he arrived at the University of Texas in Austin to work on his doctorate, he found himself caught up in the movement to bring mathematical precision, clarity, discipline, and true logic to computer programming. At that time, researchers were in the midst of developing a kind of linguistics for artificial languages, setting rules that specified what ideal computer languages ought to convey and how they could say it most effectively.

By the time he completed his dissertation and arrived at the University of Maryland in 1970, Basili was ready to design his own computer language. When his language, called SIMPL, first appeared, it attracted attention for its clarity and elegance but never became popular. For Basili, the design of a computer language was part of an inevitable progression toward an increasingly practical orientation. It led to the writing of a computer program—a compiler—that converted statements written in SIMPL into strings of digits a computer can understand. To build a reliable compiler, Basili found he had to think about the techniques one uses to write a program. Eventually, he and graduate student Joe Turner came up with an approach for developing software called "iterative enhancement."

Suppose you were building a house from scratch and you wanted to live in it as soon as possible. You also wanted to allow for design changes as the project progressed. The first step would be to build the shell of the main part of the house, including all the plumbing and electrical connections needed for it to function. This first "iteration" would probably have little more than a working bathroom and kitchen and a place to sleep, but it would be livable. Additional rooms, fixtures, and walls could be added or fixed up step by step, with each step bringing the house closer to its original specifications. Along the way, it would be possible to modify the design where necessary, because living in the house would show what worked and what didn't work.

Such an approach, applied to software, sounded good on paper and apparently worked in a demonstration project involving the development of a SIMPL compiler, but that wasn't good enough for Basili. He wasn't convinced by occasional testimonials from colleagues who had tried the method and succeeded. "I said to myself, you don't build anything without seeing whether you have done a good job. You go back and try to evaluate—measure—what you have done," Basili remembers. "It was the next obvious step for me."

It was not obvious to the rest of the computer community. In the early 1970s, very few researchers took the trouble to take the additional step of carefully, scientifically evaluating a process. Anecdotes and case studies formed the bulk of the information available on how well a particular scheme worked, especially in comparison with competing models. It wasn't at all clear what one needed to measure or how one would go about measuring it. "When we first started talking about it, people thought we were crazy," says Basili.

Measurement and experimentation have generally played at best a minor role in both computer science and software engineering. It costs a lot of money and effort to do controlled experiments, and that is too high a price for most researchers equipped to do such studies, especially in the world of large-scale software.

Moreover, as programming guru Robert L. Glass pointed out in a 1991 collection of reflective essays entitled *Software Conflict*, "There's a strange climate in the computer science and software engineering worlds, a climate in which the experimental component that is present in most other sciences and engineering is simply not present. The researchers who ought to be best qualified and most motivated to do software experimental research are simply not doing it."

Forging ahead, Basili built experiments and evaluations into the courses he taught at Maryland. At the end of each course he would conduct a postmortem, asking the students a variety of questions to get them to reflect on their projects. What do you think you did right? What did you do wrong? What would you do differently? If you started all over again, what changes would you make? Taking note of what didn't work was just as important as talking about what did.

By 1976, Basili's studies and experiments had become quite elaborate. That spring, he and student Bob Reiter decided to see whether a carefully defined, disciplined approach to computer programming had clear-cut benefits over ad hoc approaches that brought a potpourri of techniques to bear on a given problem. It was the kind of issue that could produce heated arguments among computer professionals. On one side were programmers who preferred to remain unfettered and free to create software using whatever tricks they deemed necessary. Their opponents favored structured methods, with close adherence to fundamental principles of software development.

Basili adopted a two-pronged approach to this issue. He divided his students into three categories. Some worked individually, using whatever methods came to mind. Others worked in groups of three, again using whatever they happened to think of in writing their joint computer programs. The third category consisted of three-person teams, each directed to follow a particular disciplined methodology. Basili studied the effects of team size and methodology on the resulting programs.

The students' task was to build a subset of a SIMPL compiler, using the SIMPL computer language, which required about two months' worth of effort per team. All of the participants were experienced programmers; a few had as many as three years of professional programming experience. The students were aware that they were being monitored while they worked on their programs, but they had no knowledge of what was being observed and why. In fact, most of the data for the study was collected unobtrusively and automatically each time the students tried something out on the Univac 1100 computer they used.

Basili ended up with a vast amount of data and no clear idea of what to do with all of it. But certain trends were evident, even in preliminary analyses. The disciplined teams, which spent a lot of time in planning and designing their systems before writing much of their code, required fewer computer runs, made fewer revisions, and spent less time debugging their programs than the individuals and teams using ad hoc methods. The obvious conclusion: A disciplined approach improves the efficiency of software development. Indeed, for all the factors considered in the study, the disciplined approach never appeared to get in the way of programming effectiveness or software quality.

THE CALL from Goddard's Frank McGarry came in the midst of Basili's 1976 experiment. The chance to conduct studies and experiments in a setting more realistic that the college classroom proved irresistible. By August, McGarry, Basili, and his Maryland colleague Marvin V. Zelkowitz had formed the Software Engineering Laboratory (SEL), joined also by Jerry Page and other representatives from the Computer Sciences Corporation (CSC), which held the contract to produce Goddard's software.

The SEL was a convenient label for the coordinated activities of individuals linked by a common purpose. It had no headquarters, no central location, no telephone number, and no place on any organizational chart. It was a "virtual" organization, existing only in the minds of its participants.

In the fall of 1976, Basili spent a sabbatical at Goddard, using a large part of his time to compose forms and questionnaires that programmers could fill in to report what they were doing. The collected information would presumably allow researchers and management to analyze both the software development process and the software itself, spotlighting the effects of various "improvements" on the quality of the software produced. Nothing so comprehensive had ever been done on such a large scale to assess the whole process of software development on a regular basis. It took years of additional effort to learn how best to collect, understand, and effectively use the SEL's information at Goddard.

Initially, the professional programmers at CSC, under contract to NASA, saw little value in the project and were reluctant to participate. They bemoaned the required paperwork and resented the intrusion of monitoring and other constraints into their daily work. They felt comfortable with what they were doing and firmly believed they had developed a considerable expertise in writing large computer programs.

Those attitudes began to change as the programmers realized that Basili, McGarry, and their collaborators were listening to their ideas and using the collected data on individual projects to learn about what was going on. The project information also gave them immediate feedback on how well they were doing and allowed them to have a say in the introduction of new methods for software development. Moreover, the monitoring by itself—kept as unobtrusive as possible—served to encourage better work.

One of the lessons Basili learned early on was to make sure that data collection fitted naturally into the process of designing and writing computer programs. He focused on simple questions that people could answer immediately on the forms. In Basili's scheme, no one had to do extra work. They merely had to say what they were doing at any given moment—whether reading a computer program line by line to look for errors or noting what decisions they made in going from one step to another.

He also allowed programmers to volunteer information. On several occasions while conducting audits of software development processes at various companies and institutions, he had noticed that people involved in a project more often than not were eager to provide information, but no one happened to ask the right questions or provide an appropriate means of getting at the information.

At Goddard, Basili and his coworkers spent much of the time during the first four or five years of the Software Engineering Laboratory's existence simply learning what actually goes on in developing software. They gradually built up a descriptive model of what happens step-by-step in bringing a computer program from concept and specification to final product. Once written down, this model could be refined further as additional data became available. Managers and programmers could compare ongoing projects with the model to see if they were following the correct pattern. If they detected an unexpected mismatch, they could look for potential causes and perhaps rectify an incipient problem.

"Once you know why it happens, then you can change what you're doing," Basili says. "You can change the process to avoid defects."

A history of how software development actually occurs within a given organization also makes it possible to adapt the process to the product, whether it's a microprocessor-controlled toaster or a sophisticated spacecraft. For routine projects, it may be possible to reuse large chunks of previously written programs. For unique projects, the programs would have to be developed from scratch, which entails greater risk, higher costs, special precautions, and greater scrutiny.

Over the years, Basili and his colleagues conducted several major controlled studies to look at specific issues affecting software development at Goddard. These scientific experiments were expensive and complicated endeavors, but in many cases, they provided the first reliable evidence that some approaches clearly work better than others. That's especially valuable in the field of software engineering, where opinions are often strongly held, vigorously advocated, and more prevalent than real data.

One such study concerned the most effective ways of removing errors from software. Error removal can be the most expensive part of software development, and the issue of how best to test software for errors is emotionally charged and strongly colored by opinion. Moreover, testing can't prove that all bugs have been removed. Some experts have argued that review processes—simply reading and closely examining the design, the code, and the tests themselves—tend to unearth more errors faster than does testing by itself.

To help sort out concerns about testing, Basili and colleague Richard Selby compared the combination of detailed design and code reading with two different types of testing: functional testing to see if a program met all of its requirements and structural testing to see if all parts of the program had been tested. They seeded a number of relatively short programs with various types of errors similar to those typically committed during software development. It was up to thirty two experienced professional programmers and forty-two students from the University of Maryland to find the mistakes.

On average, participants found only about half of the faults in four different programs. Interestingly, code reviews—not testing—proved the most effective means not only of identifying nearly all types of errors but also of finding them quickly. Yet, when asked which method seemed best to them, 90 percent of the participants voted for functional testing. They mistakenly believed that this type of testing had found all or nearly all of the faults. However, those who spent their time simply reading the lines of the computer program correctly perceived that they had failed to find all the errors. The process of poking around in the code to try to figure out what it does appeared to give programmers a better perspective on how well they had succeeded.

There was another curious result. In about 10 percent of the cases, programmers found a bug but didn't recognize it as one and failed to report it. Such occurrences may be more common in practice than previously realized, Basili comments.

This study, done in the early 1980s, had a tremendous impact at Goddard. Prior to the experiments, CSC programmers had typically spent nearly all of their time testing their programs to find errors and very little time on reading and reviewing the code. The convincing demonstration of how effective reviews can be in identifying faults changed all that. It countered the conventional wisdom that when it was crucial to make sure there were no bugs in a piece of software, the best course was to test it hard, then test it some more, and then test it again.

PROGRAMMERS GENERALLY FIND it very hard to resist the urge to begin coding as soon as practicable—preferably right away. Once the requirements are reasonably clear—and sometimes even before then—the programmer is writing out the first few lines of his or her program, eager to try this initial section out on the computer to see if it works. Then it's on to the next section or to a more elaborate version of the first, and so on, until the program is complete. Running the program—executing it—on a computer is built into the process of composing the software, and for many programmers, tinkering with software to get it to work, with immediate feedback, has considerable appeal.

So, when someone asks programmers to give up this trial-and-error type of software development in favor of a process in which coding represents the *final* step, in which nothing is tried out on the computer except at the very end, they naturally greet the idea with considerable skepticism and even dismay. They can't imagine dispensing with the cycles of writing and debugging that typify the process of writing computer programs. It's like asking a chef who has spent a lifetime creating new recipes by adding a pinch of this or that and tasting the result to wait until the final garnish adorns the dish to sample it.

But that's precisely what an approach known as the Cleanroom software engineering process calls for. Proponents of this approach argue that by carefully observing the method's strictures, development teams can practically eliminate errors. They dogmatically prohibit the normal types of debugging and testing techniques used in conventional programming, emphasizing instead a logical and statistical approach that puts the emphasis on mathematical verification and careful inspections. The few errors that remain— usually detected in random testing based on how the software would be used—arise out of human fallibility, out of the difficulty of manually making sure that the underlying mathematics is correct.

It's no ordinary medicine man who's hawking the magic Cleanroom elixir. An imposing figure both in stature and intellect, Harlan D. Mills speaks of the intrinsically mathematical nature of software with a calculated passion. To him, the key part of software development is not programming itself but in deciding what to do, and mathematics makes it possible to avoid errors and stay on the right track. Now retired from what was once the federal systems division of IBM, which produced software for the government, he continues to promote his ideas vigorously.

Not surprisingly, Mills was educated as a mathematician, obtaining his degrees in the early 1950s from Iowa State University. He spent some time teaching at Princeton, where a new discipline of mathematics applied to computation was beginning to emerge. From 1954 to 1957, he worked at General Electric, where he created a course for GE executives on operations research and management science, reflecting his strong interest in the application of rigorous engineering practice to business concerns. By the time he joined IBM in 1964, Mills was ready to tackle the issue of bringing greater rigor and scientific discipline to software development.

The 1960s had seen one major software fiasco after another. Several complex military intelligence systems, a huge airline reservation system, and other large projects were abandoned after years of effort and total failure. "But out of these failures, two major developments arose—first, more systematic management procedures and second, the beginnings of mathematical formulations of what previously seemed just programming ideas and folklore," Mills noted in an address he presented at a 1987 computer conference just before he retired.

The term "software engineering" first appeared in 1968, and it quickly became a buzzword. Short courses for professionals, lasting a week or less, emerged to introduce the ideas of structured programming and other proposed remedies. Unfortunately, "the ideas of structured programming were simplified to the point of easy listening rather than substantial study," Mills declares. The missing element was the mathematics that would enable a programmer to prove that the statements of a program actually did what the requirements specified. Ironically, mathematics of this type was just as scary and intimidating to programmers and engineers as mathematics in general remains to much of the populace.

Victor Basili first encountered Mills in the early 1970s. The two eventually ended up working together in presenting occasional seminars at the University of Maryland, where Mills sometimes taught classes. It was from Mills that Basili received strong support for the notion that carefully "reading" a computer program may be more effective than testing for finding and eliminating errors, and Mills himself developed techniques to make code perusal more systematic, thorough, and convenient, especially for large, complex programs.

Mills was also a strong advocate of what he called the "chief programmer" model of software development, in which different people on a team take primary responsibility for different aspects of a project. Basili adopted a similar structure in his experiments involving the value of disciplined approaches to creating software. The need for specialization in software design and development has since become an important part of Basili's campaign to improve computer programming.

Basili's experience a few years ago in observing the design and construction of an addition to his house reinforced this belief. He noted, for example, that on a building project, no one would ever ask the architect to do the plumbing, carpentry, or electrical work. Appropriate specialists handled different aspects of the project. In software development, however, programmers are often expected to handle everything that it takes to build a working computer program—and nowadays, some of these programs are far more complicated to construct than a skyscraper, let alone an addition to a house.

"Part of the reason we don't recognize this is that people still don't think software [development] is hard," Basili ruefully notes. Software developers are expected to be magicians capable of incredible juggling acts.

By emphasizing the notion of computer programming as a fundamentally mathematical pursuit, Mills and several collaborators gradually developed an entire system for producing practically defect-free software. It's a purely mathematical, tightly controlled style of programming, which begins with clear statements of overall requirements. These specifications are then carefully and precisely broken down into subspecifications over and over again until the actual statements of the program, written in a given computer language, are finally reached.

In some sense, the process is analogous to writing a novel by plotting the entire manuscript before writing the individual sentences. The author starts with a brief outline, sketching the major events as they occur from beginning to end. Then each event is elaborated, with key details added chapter by chapter. The details themselves are worked out further, and so on, down to the individual events, characters, language, and other ingredients that go into the story paragraph by paragraph. Finally, the author is ready to compose the sentences to put his or her plan into effect.

Of course, novelists don't usually work this way. Some start writing immediately, and some spend inordinate amounts of time composing the first sentence of the first paragraph of the first chapter, before building the story step-by-step. While writing, an author is often editing, fixing, and refining—even changing the plot or going back to fill in particulars crucial for subsequent plot twists. Indeed, what seems natural for many novelists feels right for many programmers, and they also plunge in and quickly get caught up in their cycles of writing and debugging. What makes software development a different sort of art is that every statement counts, and errors can matter a great deal.

Initially, the notions upon which Mills and his colleagues built their Cleanroom methodology proved too radical, austere, and different from conventional practice for managers at IBM. It was not until 1987 that Mills and his group were finally allowed a chance to try out the full scheme on a sizable project. They used their techniques to develop the software for a military helicopter flight avionics system. Completed ahead of schedule, the program was a modest 33,000 lines long. Final testing revealed an average of 2.3 errors per thousand lines, less than one-fifth the rate commonly found in software developed using more conventional methods.

A year later, the team produced IBM's first commercial product developed entirely using Cleanroom methodology. It was a program of 85,000 lines for reorganizing programs in a widely used computer language known as COBOL to make them more efficient and understandable. Again, relatively few errors were found during testing, and none was discovered after the product went to customers.

By 1990, there were enough successes that IBM managers started to take notice, and Cleanroom software engineering began to spread within the company. Three years later, the company was offering for sale computer programs that aided professionals interested in using parts of the Cleanroom methodology. At the same time, Mills and others presented numerous seminars, workshops, and tutorials to disseminate their concept, joining the hordes of medicine men and women promoting particular cures for endemic software ailments.

And the ailments certainly persisted. Mills himself liked to shock his audiences by observing that perhaps as much as half of the software commissioned and paid for by the federal government never sees the light of day, often because the programs prove unworkable, are completed late, or turn out to be faulty. That's a serious drain on the federal budget because the government purchases more software, worth at least $100 billion annually, than any other single buyer in the United States.

Even the vaunted IBM federal systems division (now owned by Loral), which had produced the unusually high-quality software for NASA's space shuttle (see Chapter 2), suffered serious setbacks in developing a new, sophisticated air traffic control system for the Federal Aviation Administration. Far behind schedule and over budget, the project in 1994 still looked as if it would never meet its goals.

However, it's one thing to have the inventor of a system using it in a controlled setting and another for someone initially unfamiliar with the process to try it out. Would Cleanroom techniques really work for programmers other than Mills and his colleagues? Would they work for computer programs that are longer and more complex than those in the initial tests of the method?

Such questions were not lost on Basili and the software engineers at Goddard. In 1988, they embarked on their first controlled experiment involving the Cleanroom method. At that time, the method had garnered a considerable amount of publicity, but there were few demonstrations of its efficacy.

Basili started modestly. His experiment involved fifteen three-member teams of advanced students at the University of Maryland, ten of which used Cleanroom methods while the other five used a more traditional approach. Their assignment was to develop software for an electronic message system. The resulting programs were relatively small, averaging about fifteen hundred lines.

All of the Cleanroom teams were able to build their programs without having to use the computer to try them out during development. They relied on manually checking the programs by reading through them to look for faults and omissions. In the end, these teams succeeded in producing the software on schedule and with fewer errors than those using conventional approaches. Although most of the participants using Cleanroom methodology commented that they would use it again, some of them said they missed sitting in front a computer screen to watch how well a program or section of a program worked while they were in the midst of developing it.

The fact that the Cleanroom process proved effective for a small product didn't necessarily mean that it would work on a larger scale. To use the mathematical methods at the heart of Cleanroom software engineering required special and extensive training for the programmer. Mills himself long maintained that learning the mathematical basis of software engineering requires four years of undergraduate study, not the few hours or days of a typical workshop or tutorial.

The need for such training became evident in subsequent experiments involving larger projects at Goddard. At one point, Mills came up from his Florida home to help teach Cleanroom methodology to the programmers involved in one of the experiments. The additional training helped, but the results were ambiguous. It wasn't always clear that the Cleanroom technique would actually improve productivity, at least not without a significant investment in education.

Basili and his colleagues now tentatively conclude that key elements of the Cleanroom approach can be used at Goddard, especially for projects involving fewer than fifty thousand lines of code. Although they don't see similar reliability and cost gains for larger efforts, they continue to investigate the method to sort out what works from what doesn't.

THE CLEANROOM EXPERIMENTS represented only a small part of the activities conducted under the umbrella of the Software Engineering Laboratory. From the time of the laboratory's inception in 1976, Goddard's flight dynamics division completed more than a hundred software projects, with six to ten projects going on at any given time. This software—a total of at least 4.5 million lines of code—supported more than twenty-five NASA missions, including the operation of the *Cosmic Background Explorer*, the *Gamma Ray Observatory*, the *Extreme Ultraviolet Explorer*, and other Earth-orbiting satellites.

During this time, typical programs more than doubled in size, reflecting the increasingly sophisticated demands placed on controlling spacecraft. Yet by learning from their successes and failures, the software engineers at Goddard and CSC could produce the considerably more complex software required in the 1990s at slightly less than it cost to produce comparable software in the 1970s. At the same time, the various studies and experiments conducted at Goddard and at the University of Maryland yielded more than 250 documents and reports.

The SEL approach gradually evolved into a system for packaging and reusing experience. It was no longer enough just to get a computer program out the door and into use as soon as possible. Constant, comprehensive monitoring of every software project provided up-to-the-minute information on how well a project was going, allowing time to make any necessary adjustments, and these data became part of the extensive, growing repository of information against which new initiatives could be compared.

Basili calls this system for packaging software engineering expertise the "experience factory." It has become the centerpiece of his current missionary efforts to improve software quality.

In October 1992, Basili found himself in a small television studio at the University of Maryland. He had finally been able to squeeze in the time to present before the cameras a two-day course that highlighted his "experience factory" concept. Broadcast live by satellite to a handful of sites across the country and recorded on videotape for subsequent editing and packaging into an educational television series, his twelve-hour marathon of lectures provided him with the opportunity to distill his own experiences into a coherent argument on how to improve software quality. It gave him a chance to look back on more than two decades of work spent trying to get things right.

Basili began by reflecting on the special nature of software— the characteristics that set it apart from nearly all other manufactured or crafted products, such as toasters or cars. Automotive engineers, for example, can spend years designing a car, but at some stage, the design goes into production, and from then on, industrial engineers handle how best to manufacture the product, how to assure its quality, and how to build large numbers at the lowest possible cost. They deal with metals and machines, with nuts and bolts and paint, with the mechanics of assembly lines. Software, on the other hand, is almost entirely design and development. It is a product of the mind, and it remains so throughout its lifetime. This distinctive characteristic sets software apart from other products.

Moreover, engineers generally have a much better feel for what a toaster or a car should look like and how it should operate than what a computer program should look like and how it should function. When automotive engineers design a vehicle, they can readily see, for instance, that adding a fifth wheel to the family car would be silly. They can picture the result, and it obviously makes no sense. They can rely on their intuition to judge whether something will work. But in dealing with software, engineers and programmers have very little to go on when deciding what makes sense and what doesn't in any given situation. Practically anything is possible, and it's no trouble at all to add the unnecessary fifth wheel.

Furthermore, a software product is never really finished. It can always be changed easily and quickly. "We don't understand the meaning of change—what can and cannot be done," Basili says. In a car, engineers can sense that it's far easier to alter the arrangement of indicator lights on the dashboard than to move the steering column or add an extra wheel. With software, no one really knows. At the same time, "we make a difficult problem more difficult because we often have unclear goals and requirements," Basili remarks. There's a sense that because software can always be modified, any problems that may come up along the way can be fixed later. At the same time, "we don't treat [software] with respect," he declares.

Supposing a materials scientist came to the president of an aircraft company with a new, revolutionary metal alloy perfect for manufacturing lightweight airliners and insisted that the metal be introduced to the production line the next day. The scientist would be taken away in a straitjacket for such an outrageous recommendation. The company would want to experiment with the metal first, testing it on a small scale, then gradually expanding its use if the experiments proved successful. Immediate adoption would be out of the question.

In contrast, people expect similarly radical changes to occur routinely and quickly in the software domain. Suppose that a software engineer announces that he or she has developed a novel strategy for producing defect-free software or that a company manager attends a one-day workshop featuring a new, widely touted methodology that allegedly increases software productivity. More often than not, the result is a company-wide or division-wide edict to adopt the new technology. There's often little training, no testing, no gradual implementation, no evaluation to determine where to use the methodology, how best to use it, or what its limits are.

The manager or software engineer earns a reputation as a great innovator. People who resent or oppose the imposition of the "innovation" get labeled as being "resistant to change." And they get blamed for its inevitable failure.

"That's not innovation; that's stupidity," Basili retorts. Yet, it happens again and again in the software domain, costing companies, agencies, and other organizations a great deal of money, experience, and stability.

What Basili has learned in his years with the Software Engineering Laboratory is that it takes time to get things right. "We learned through a great deal of pain," Basili says. "We all thought when we started that we'd have it wrapped up in a few years."

"The good news is that although it has taken us a long time, I think we've gotten to where we understand what the problems, are," he continues. "We understand the things people have to do. We also know how to do them, though they're hard." As a result, programmers in the flight dynamics division at Goddard have greater control over what they do than programmers at many other organizations.

But would the experience factory approach—with all its measurement paraphernalia—work elsewhere? Basili readily concedes that measurement and model building by themselves won't solve problems in software development. These techniques should augment but not replace good management and engineering judgment, qualities that are sometimes in short supply. So he's scrupulously cautious about judging whether the same process would improve the situation in other settings, which may differ enormously from Goddard.

Like Basili, Goddard's Frank McGarry recognizes that particular results from specific software technologies may not necessarily apply in other settings. However, he believes that the process by which an organization can analyze its software strengths and weaknesses and thereby benefit from that information in finding the appropriate techniques for its needs is exportable. NASA itself has shown interest in extending some form of the experience factory concept to other centers. In fact, some centers such as the Jet Propulsion Laboratory, which operated the *Voyager* missions to the outer planets, have already worked out their own, systematic methods for improving software quality.

**WHILE THE EXPERIENCE FACTORY** concentrates on techniques for improving an organization's software development process on a case-by-case basis, the Software Engineering Institute (SEI) offers a different, broader approach to improving software quality. In an elaborate, somewhat bureaucratic way, it defines the optimal qualities that a generic software development process ought to have and provides a means of checking to see how well a particular organization stacks up against this standard.

Ensconced in a glass and concrete fortress in Pittsburgh, the SEI originally arose out of severe software problems at the Department of Defense. Software errors, in particular, have long been a special concern at the Department of Defense because almost every system in the current and planned military inventory relies extensively on computers and microprocessors. For example, computers control the targeting and flight of missiles, coordinate and control sophisticated systems installed in high-performance aircraft, and integrate the complex activities of battlefield command. They play key roles in training. Military planners foresee that an entirely new form of warfare may evolve, based on the use of computers and methods to attack, deceive, and neutralize them. Consequently, software has become a critical element of military systems.

Within the Department of Defense, software development "ranges from a reasonably effective, disciplined approach in a few systems to near chaos in others," Edith W. Martin, former deputy undersecretary of defense for advanced technology, reported in 1982. A U.S. Navy study, for example, revealed thirteen different mathematical algorithms, or formulas, in use for steering an air-plane from one place to another. In the U.S. Air Force, up to 90 percent of the programs were coded in a primitive, difficult-to-decipher computer language rarely used elsewhere. A U.S. Army survey of about one hundred battlefield systems revealed the use of thirty-four different versions of essentially the same computer, each operated by software written in different computer languages. Such diversity created headaches for those responsible for testing and upgrading the systems, and it produced problems on the battle-field when one computer had to communicate with another.

Situations like this led the Department of Defense to launch in the spring of 1983 an initiative, called STARS (Software Technology for Adaptable, Reliable Systems). One aim of STARS was to create a software engineering institute where the department, in cooperation with industry and universities, could evaluate and demonstrate the usefulness of new programming techniques and integrate those ideas into military systems more quickly. The institute was also charged with training department personnel and sifting through the literature to spotlight useful techniques and innovations. Plenty of programming and testing tools existed, but this information was often hidden in obscure journals, locked away in computer testing centers, or scattered in bits and pieces and applicable only to particular computers and computer languages.

One of the SEI's key initiatives was a program for scrutinizing and controlling the process of developing software. Led by Watts S. Humphrey, who had worked at IBM from 1959 to 1986, the institute developed what it called a "capability maturity model." Companies could obtain questionnaires to perform self-assessments of their software development prowess, while Department of Defense personnel could use a similar yardstick to evaluate potential defense contractors offering to do software development.

The model defines five levels of maturity, ranging from chaotic (level 1) to well-organized and self-improving (level 5). In going through this evaluation to determine how they rated, 80 percent of organizations found themselves in level 1. Only 12 percent had reached level 2, and 7 percent level 3. The figures were even worse when the evaluation was applied to individual software projects. Of these, 88 percent fell into level 1. Even the few organizations claiming to have projects that attained a level-5 rating in these self-administered evaluations admitted their software development processes had room for improvement.

The problem with such generic standards is that they are descriptive rather than prescriptive. There is still no way to guarantee that by doing the things required to reach a higher level of maturity, one will end up producing software of higher quality. Moreover, a few companies have accused the Department of Defense of misusing such evaluations to favor some contractors over others. And there's no guarantee that the military evaluators themselves are well qualified or do the job fairly.

Even at the SEI, researchers are beginning to focus on the human variable in the elusive quality equation. They are starting to recognize that psychological barriers to improvement may be far more important than technological barriers. Indeed, there's a tremendous gap between what can be done and what is actually done, and a great deal hinges on the quality of the personnel in any given organization. There just aren't enough highly qualified, experienced, knowledgeable people to get the jobs done properly.

From Basili's point of view, human factors are inextricably linked to improving software development. Improvements in quality and productivity must come slowly to make sure they take hold. Introducing new ideas requires considerable patience, tact, and care. But such an approach runs counter to the rapid pace at which computer technology is evolving and the burgeoning demands for increasingly sophisticated software to operate this machinery.

"All of this technology is really human-based," Basili says. "We're doing development, and that means people have to change. What we understand in our society today is how hard that is. Change is occurring, but it can happen only very slowly. If you change too fast, you lose total control."

In 1983, at a software test and evaluation conference held in Washington, D.C., Basili concluded his presentation by saying, "It is almost frightening how many open questions there are in a field where we have been working so long." More than a decade later, the situation seems hardly improved.

But Basili remains optimistic. "We always seem to be pushing the boundaries," he says. "But I have a basic belief that people are smart. Of course, we'll get into trouble. We may even have a couple of major accidents first. But we'll step back; we'll learn."

It's a cautious sort of optimism.

# CHAPTER 5

# Time Bomb

---

**PAST SHELVES CRAMMED** with candy bars, cosmetics, and headache remedies, beyond racks of toothbrushes, sunglasses, and greeting cards, behind a high, gleaming, fluorescently lit counter, the pharmacist plies an ancient trade. Against a backdrop of tall cabinets lined with jars of capsules, lozenges, tablets, and treacly liquids, he or she quietly taps away on the keyboard of the indispensable pharmacy computer. Every prescription passes via the keyboard into a computer-mediated world of elaborate record keeping, inventory control, and label printing.

But when the pharmacist types in June 18, 1899, as a customer's birth date, the computer responds with the message: INVALID ENTRY. In a manipulation invisible to the pharmacist, the computer's program actually registers only the last two digits of the year. So when the computer calculates the customer's age, it simply subtracts 99 from the two digits representing the current year and comes up with a negative number. Because this result is unexpected, the computer rejects the typed data, much to the pharmacist's dismay. The programmers responsible for the software had somehow overlooked the fact that a pharmacist may occasionally deal with customers born before 1900.

Computer programs that rely on just two digits to mark the year have operated erroneously in other situations. In 1992, for example, a 104-year-old woman in Minnesota received an invitation to attend kindergarten because school officials had instructed a computer to search a central database for the names of people born in '88. Similar date problems once plagued record-keeping software used in hospitals, so many hospitals have switched to software incorporating three-digit years, dropping only the first digit (because it is always "1") to save storage space. This fix, however, will stop working when the year 2000 arrives.

Much of the software used today in business and industry still registers dates in six-digit or five-digit units. For example, September 14, 1987, appears as 870914 (two digits each for the year, month, and day) or as 87257 (a two-digit year followed by the day of the year). In the early days of data processing, these formats were particularly useful for sorting data by date to arrange them in chronological order. Such units could also be used for calculating time spans—whether someone's age or the number of years left for paying off a mortgage—or as identifiers to keep track of stock market transactions, customer orders, and other dealings. Enshrined in federal government standards in the 1960s, this convenient format was perpetuated in program after program.

Problems involving dates—and information computed on the basis of dates—have already surfaced, especially in software written years ago but still in use. In many instances, the shortcuts that programmers once took to speed up a procedure or to save storage space in the computer's memory are no longer necessary, but such simple subterfuges as using only two digits to represent years have survived in software enjoying a lifetime much longer than its developers had originally envisioned.

The biggest headaches could come with the arrival of midnight at the start of the first day of the year 2000, when many of today's computer programs will still be in use. "There'll be a major crisis," predicts Elliot J. Chikofsky, a computer consultant and lecturer at Northeastern University in Boston. "If we're not careful, it'll be algorithmic anarchy."

Consider Monday, January 3, 2000. In the six-digit format, this date would appear as 000103, and December 31, 1999 would appear as 991231. Which is the later date: 991231 or 000103? In the absence of other instructions, the computer will of course decide that 991231 is larger than 000103, and December 31, 1999 suddenly becomes the later date.

In the year 2000, "any system that stores less than the full four digits of the year must, for the first time, deal with a year number that is smaller than its predecessor," Chikofsky notes. "Comparisons based on inequality will suddenly change direction. Subtractions to discover [the] time interval will yield a negative number."

Because it typically takes more than a decade for new technologies to reach sufficiently high standards of usability, reliability, and quality for widespread industrial and commercial use, the information infrastructure for the year 2000 is already in place. Dates play crucial roles in networks of automated teller machines, telephone verification of credit-card sales, insurance databases, airline reservation systems, and other components of today's information-dependent society. Many of these transactions are conducted automatically, with no human in the loop to check them for sanity.

For systems relying on fewer than four digits to designate the year, the consequences in the year 2000 may very well produce a mind-bending time warp of catastrophic proportions. In the financial world alone, the results could include the scrambling of interest calculations, delays in pension benefits, misrecorded loan payments, and unwarranted foreclosure notices. "Imagine making a phone call to another time zone at midnight [a happy New Year greeting to a relative, perhaps] and being billed for ninety-nine years," Chikofsky adds.

Averting this impending disaster requires considerably more effort than merely switching to four-figure years. Many computer programs use dates for a variety of purposes—from calculating time spans to determining whether one version of a computer file is older than another, then automatically erasing or overwriting the older one. Finding and fixing all of these instances—some deeply embedded in the software—is no simple matter.

"Every routine that directly or indirectly depends on a date is suspect," Chikofsky warns. "Does it subtract to reach a conclusion? What happens if it gets passed a bad date? We must check any place a date is used in *any* way, including birth dates, transaction dates, and files named by encoding their date and time."

Problems involving dates and time go beyond the unfortunate use of two-digit years. Indeed, synchronizing human time with computer time presents programmers with a host of difficulties. Although computers hum to a precisely defined, inflexible rhythm that sets the pace of games, database queries, and calculations, they must also function in the context of a calendar plagued with the quirkiness and arbitrariness of human experience and thought— sixty-second minutes, sixty-minute hours, twenty-four-hour days, switches between standard time and daylight savings time, seven-day weeks, months of varying length, and leap-day and leap-second corrections to bring the calendar into accord with celestial time.

From a computer's point of view, such quaint conventions are at best a nuisance and at worst a nightmare. Every four years, for example, computers of all sorts fail because unwary software developers forgot that February has twenty-nine rather than twenty-eight days. In 1992, for example, legions of ATM machines failed to acknowledge the extra day and ended up scrambling the data encoded on cards of customers who used an ATM on February 29. An electronic-mail system crashed because the software couldn't recognize the date affixed to messages handled by the system. In Iowa, all liquor licenses expired on February 28, and new licenses didn't come into effect until March 1. State officials were forced to announce that this was due to a "computer error" and promised not to enforce the law for establishments caught by the leap-day glitch.

Such occurrences provide a seemingly endless supply of material for Peter Neumann's Risks forum (see Chapter 1). "You might think that the leap-day problems would have been adequately anticipated, especially in the four years since their previous incarnation," Neumann noted in a 1992 column in *Communications of the ACM*. "Getting clock arithmetic correct might seem to be a conceptually simple task—which is why it is not taken seriously enough. But even if earlier leap-year problems were caught in older systems, they continue to recur in newer systems. So, now we have four more years to develop new software with old leap-year bugs, and perhaps even some creative new ones!"

The year 2000, however, brings an additional twist to the leap-year saga. To keep the calendar synchronized to the movement of Earth around the sun, leap years occur every four years, except in three out of four years ending in 00. For these centesimal years, only those evenly divisible by 400 are leap years. Thus, 1600 was a leap year, while 1700, 1800, and 1900 were not. The year 2000 will be a leap year. When Chikofsky gives lectures to computer professionals, he likes to have them vote on whether the year 2000 will have a February 29. Invariably, the response is divided.

LIKE CHIKOFSKY, other observers of the computer scene have warned of the potentially disrupting consequences of the widespread use of two-digit years in software. As the year 2000 approaches, attention to the problem is steadily increasing. But Chikofsky brings a special perspective to the subject because he's a strong proponent of a new software engineering specialty called reverse software engineering. It involves the development of automated techniques for recognizing what a computer program does—for recovering information from existing software.

Indeed, the turn-of-the-century date problem pales in comparison with the difficulties that businesses and other users of antiquated software already confront in trying to change the way they handle data and operate their enterprises. Companies face the daunting prospect of deciphering enormous computer programs, often written more than a decade earlier. More often than not, the original programmers have changed jobs, and their knowledge of the program's internal workings and the reasons why certain things were done have disappeared with them. Meanwhile, other programmers may have patched up or modified the program, adding to the mess.

The notion of reverse engineering goes back to the days when competitors would carefully take apart a rival's product—a car, a toaster, or an integrated-circuit chip, for example—to learn what makes it tick. The practice remains commonplace, but it's done primarily to reconstruct an existing product for which no specifications are available.

Reverse engineering of software differs from that of hardware in that it is usually applied not to a competitor's product but to someone's own computer programs. The oldest and most deeply entrenched computer programs are often described as legacy systems. Analysis of this archaic, though frequently vital, software to rediscover what it really does and how it does it constitutes a crucial first step toward making changes in or replacing such a system.

For example, suppose that a bank's computer program for calculating interest, printing out customer statements, and performing a variety of other functions consists of a million lines of instructions (or code). When new tax laws suddenly force the bank to report interest earned on accounts in a different way, updating the software can turn into a nightmarish task. Programmers must examine the million lines of code to figure out which parts are relevant, make the necessary changes, and ensure that the new capabilities don't adversely affect the program's existing functions. Reverse-engineering techniques supply some of the necessary information.

A vast number of companies introduced computers into their operations decades ago, and many still use the original programs. To keep pace with changing computer technologies and business requirements, the companies have to modify and update their archaic systems or develop or purchase new software. To do either is costly and time-consuming.

The U.S. government faces an equally dire situation. In 1983, J. Peter Grace, chairman of a White House panel on cutting the cost of government, noted, "There are over nineteen thousand computers . . . in the federal government, and [both hardware and software] are, by and large, twice the age . . . as those in the private sector. And they're obsolete. In fact, they're so obsolete that in some cases federal employees have to maintain these computers because the manufacturers have discontinued servicing them. That's how bad it is."

The passage of more than a decade hasn't significantly improved the situation, and modernization efforts at the Internal Revenue Service, the Federal Aviation Administration (FAA), and other agencies and departments have foundered over the tremendous difficulties of deciphering old software to see how it should be fixed or modified.

For example, when software engineers at IBM's Federal Systems Division (now owned by Loral) received their contract to update the FAA's aging national air traffic control system, they had to rely primarily on examining the existing software itself to rediscover how the huge program's modules fitted together and how data were organized and handled. Only then could they begin to draw up an overall design for the old system from which a new system could be derived and developed.

To cite another example, the U.S. Department of Defense maintains a huge inventory of nonmilitary software, mainly for accounting, payroll, and personnel data systems. These 1.4 billion lines of code are spread out among 1,700 data centers, and they represent a great deal of duplication of resources, largely because the Army, Navy, Air Force, Marine Corps, Joint Chiefs of Staff, and various agencies developed their own systems. Up to thirty years old, many of these computer installations have poorly documented software,

and it's often difficult to obtain accurate organization-wide information because of the resulting diversity and obscurity. Already
spending about $9 billion per year to operate these systems, the
Department of Defense now faces the immense task of analyzing
what is going on in the hope of eventually consolidating the entire
operation.

In many cases, legacy software has become so unstable that information systems personnel are almost literally afraid to touch it
for fear of making problems worse. Ironically, the very flexibility
that makes software such an attractive business tool can lead to a
situation in which change becomes very difficult. An information
system can end up locking in the way the company does business.

In 1992, for example, the Travelers Insurance Corporation paid
$64,500 in fines for errors by several of its subsidiaries in the pric-
ing of automobile and homeowners' policies. The violations involved overcharging and undercharging customers for insurance.
The problems were partly caused by software faults that produced
discrepancies between rates quoted by agents and those actually
charged when policies were issued. Agents who were aware of the
problems had to work out stratagems to bypass the company's computers, giving customers an excuse such as, "We can't change the
computer."

A company's inventory of existing software also represents a significant asset. Using reverse engineering, it's often worth the effort required to analyze a given system to see what's worth preserving
and what parts should be thrown away or replaced.

Chikofsky can make the problems faced by companies, government agencies, and other computer-dependent institutions
sound both amusing and disheartening, or even tragic. Making the
situation even worse is the fact that computers increasingly work
together in networks as distributed systems. The software can be
physically stored in one computer, but other computers scattered
across the country or throughout the world may be using it, sharing it, and perhaps even modifying it at any given moment. "We're

no longer alone with our code," Chikofsky says. "In many cases, we don't even know *where* the software is." Another commentator once defined a distributed system as "one in which the failure of a computer you didn't even know existed can render your own computer unusable."

In a sense, what Chikofsky talks about is software archaeology— digging into fossil code to try to understand what it's all about. Although software dinosaurs are still alive and functioning, in many cases no one really knows how they work because there are no accurate descriptions—no documentation—of the software as it exists. In other words, the manuals are missing, incomplete, or incomprehensible.

The software situation is roughly analogous to what happens when a builder completes a house but fails to pass the blueprints on to the homeowner. To put in a new electrical outlet, the owner has to figure out where the studs are, where the wires go, and which spots to avoid. This often requires testing the walls and drilling inspection holes. There's always some risk of causing unintended damage, which must then be repaired. Computer professionals responsible for modifying software are in a similar situation, except that they're more likely to be dealing with a level of complexity analogous to the towering World Trade Center rather than to a modest home.

Even having the blueprints on hand might not be enough. The builder might have neglected to follow the blueprints, instead making modifications during construction without changing the plans. The same thing happens with software. "We can't trust the documentation even when we have it," Chikofsky declares.

The idea of examining software to determine what it does is really nothing new. Harlan Mills, Victor Basili, and other key figures in the software engineering community have long advocated the benefits of "reading" code to track down errors (see Chapter 4). In an informal sense, such examinations have always been a part of writing computer programs. Experienced programmers, in particular, create and carry around in their minds conceptual models of their programs—outlines that function somewhat like plot synopses of novels. They can then compare this model of what ought to happen with what the lines of code, as written, actually do. Indeed, it's not unusual to see a programmer fumbling with a hefty stack of sheets, fingers interlaced with the pages, to follow the flow of a program's logic.

"It is important to remember that reverse engineering does not inherently require automated tools," Chikofsky and several colleagues noted in a 1993 report on the status of reverse software engineering. "It is a process of recognition and cognition that is in play every time a programmer reads someone else's code. It is a natural part of software development and maintenance, and thousands of programmers do it every day."

Faced with an undocumented hundred-thousand-line program, a reviewer could begin by first studying the program's physical appearance—looking for compact blocks of code printed out on sheets of paper that, like chapters in a novel, each typically encompass a single, major idea or event. One can then look for expressions that recur frequently in certain passages to obtain some inkling of what these sections are for. One can look at how different pieces are linked together, ascertaining what is connected to what and what causes what. From such clues, one can begin building a mental image—a model—of what the program does. Data names provide additional evidence, and so on, until one can understand the software in an abstract sense at levels higher than the lines of code themselves. Such "reading" of unfamiliar code serves as an exercise in map making—as a recording of the results of explorations at various levels of detail.

That's the kind of detective work that reverse engineering involves: identifying the components of an existing software system and establishing the relationships among them. It's possible then to create various high-level descriptions of the program, often visualized as diagrams with blocks joined by lines in a network of dependencies and functions. It's also possible to ferret out the overarching rules built into a given system, for example, the formula by which a company computes what it charges for various insurance policies.

**CHIKOFSKY'S INVOLVEMENT** with computers began in the early 1970s as an undergraduate at the University of Michigan in Ann Arbor. He had arrived at college with a strong interest in the social sciences, but he was also an avid reader of science fiction. His roommate, who was studying computer programming, noticed the subject matter of the novels lining Chikofsky's bookshelf and dared him to take a look at the real thing by taking a computer class. In the second semester of his freshman year, with some trepidation, Chikofsky signed up for a course in programming.

He was delighted to discover that he had a knack for composing computer programs. Programming also gave him an outlet for his interest in games and their mathematical formulation, and he quickly found that he was not alone in his enthusiasm. It didn't hurt that his instructor's name was Hal, just like the ominously insistent computer, HAL, who plays a key role in Arthur C. Clarke's *2001: A Space Odyssey.*

By his junior year, Chikofsky had decided to major in computer science, which took him uncomfortably far from his original inclination toward studying political science and sociology. He spent hours hanging around the university's computer center, where students who were serious about programming tended to congregate to be near the large, mainframe computers. Affable and outgoing, Chikofsky enjoyed helping others debug their programs, even when the programs were written in computer languages he didn't know. "Once you have some idea of the basic principles of debugging, you see things—patterns—even if you don't know the language you're working in," he notes.

Although the notion of structured programming came up in

classes, Chikofsky—like nearly all other students designing games, creating databases for various purposes, or struggling to get their assignments done on time—belonged to the "unstructured" school. It was write, test, debug, over and over again, with little regard for the careful design and planning required by a structured approach. Years later, he found it tough but necessary to break old habits and take a more orderly approach to software development.

By the 1970s, a small number of programmers and researchers had begun to look seriously at automating the process of designing and writing computer programs. These early efforts represented the beginning of computer-aided software engineering (CASE), and the development of computer programs, or tools, specifically designed to assist programmers. It was ironic that it had taken computer professionals considerably longer than others to recognize the potential advantages of using their own technology to assist them in software development (though it's also possible that they were more aware than most of how error-prone such an enterprise could be).

The Information Systems Development and Optimization System (ISDOS) project, led by Daniel Teichroew of the University of Michigan, was a pioneering CASE effort. Its goal was to make specifications—the comprehensive list of definitions of what a particular computer program is supposed to do—easier to produce and use. Thus, anyone needing to know the function of a certain part of the system could simply consult the specifications, either to construct the software in the first place or later to understand the working system.

In previous decades, people often hadn't bothered to specify programs at all, and some programmers kept the schemes they had in mind entirely in their heads. Moreover, many subscribed to the idea that the program defined its own behavior. Where specifications existed, they were generally too vague, incomplete, or inflexible, and couldn't keep up with the pace of changes to the software. As a result they became obsolete with startling rapidity.

The ISDOS project involved the development of software that would help solve the problem of missing, incomplete, or ambiguous specifications and documentation that so often stymied system development and modification. It tackled the process by which analysts painstakingly collect data—conducting interviews with experts and users, and going to records and other information sources—to elucidate how an existing system functions and to de termine the requirements for a new system. In the past, designers had collated, analyzed, and summarized these data manually to come up with a set of precise specifications for building the new, improved software. The ISDOS approach provided a systematic way of expressing the collected information in a format suitable for analysis by computer. In effect, the ISDOS software handled the tedious clerical tasks associated with writing documentation, leaving the more important, conceptual tasks to the designers and programmers.

The ISDOS project had two main components. The collected data were expressed in a special language, called the Program Statement Language (PSL), as they were added one by one to a system database. This notational language, geared to expressing relationships between certain "objects," was carefully designed and structured to avoid the ambiguities of English. "Natural languages are not precise enough to describe systems," Teichroew said in 1983, "and different readers may interpret a sentence in different ways." The use of PSL avoided this problem.

A set of programs working together as the Problem Statement Analyzer (PSA) checked the information as it was entered, searching for inconsistencies, identifying errors and missing information, and sounding warnings. The result was an up-to-date documentation database that served as a kind of system encyclopedia. It contained all the available, relevant information about what happens and why it happens in the logic of the computer system under development. Shared by everyone involved with the computer system, the package made it easier to detect omissions and correct errors. At the same time, the organization using the computer would know what system data it had. It wouldn't have to depend on the memories of individuals, who were not always available when a specific item of information about the system was needed.

Analyzing an array of PSL statements, the PSA portion of the system could also generate reports, ranging from directories, indexes, and statistical summaries to representations of information flows or of relationships among different parts of the program being developed. Thus, system designers could obtain a variety of snapshots of their work in progress—from portraits conveying the overall sweep of the landscape to micrographs showing intimately detailed structures.

Chikofsky encountered the ISDOS project during his final term as an undergraduate, when he took a course on information systems design. He next went to business school but didn't like it and left after only six months. He then turned to industrial engineering, in large part because of his growing interest in the ISDOS project. By the time he completed the requirements for his master's degree, he was so busy with its work that he joined the ISDOS research staff full time, instead of going on to obtain his doctorate.

ISDOS itself had proved not only a rewarding academic research project but also a commercial success. Its PSL/PSA software had been sold to a broad spectrum of companies, including banks, aerospace firms, defense system contractors, and even major computer suppliers such as IBM.

Chikofsky's main function was to institute, and then oversee, the customer-service side of the ISDOS operation. That entailed supervising a diverse staff of students and faculty members charged with helping customers with problems, keeping documentation up to date, making improvements, and adding new capabilities to the software. It was all part of a peculiar process known in the software world as "maintenance."

Software maintenance differs from a mechanic's job of lubricating moving parts and replacing worn components on a mechanical device. Once installed and in use, software doesn't wear out physically. But software's curse of infinite flexibility allows a never-ending regimen of alterations to correct errors, improve performance, add new features, and enhance or modify capabilities. Because software is typically in a continual state of change, a program presents a moving target for anyone concerned about documenting what it does.

For example, a typical program used to calculate insurance premiums may have started out more than a decade ago, when it was designed and written for a particular type of computer. Over the years, the program was adjusted to run on other, newer types of computers with less stringent memory requirements and different operating systems. Other patches to the software introduced additional capabilities or corrected problems that had surfaced.

After such a history of haphazard changes, even the original authors of the program would have trouble recognizing their creation and figuring out how the latest version works. Of course, such a history would also have a devastating effect on a program's documentation. The result is often notebooks full of cryptic, laconic comments noting changes to the software or, alternatively, sets of seldom-updated specifications, all destined to sit on bookshelves gathering dust.

Software maintenance is also a costly operation. In 1987, the U.S. Air Force allocated $85 million for software development for its F-16 fighter jet. At the same time, the Air Force forecast software maintenance expenditures totaling about $250 million, almost three times the development cost of the original software.

In maintaining the ISDOS programs, however, Chikofsky faced an additional problem peculiar to a university environment. The staff, made up largely of students, changed regularly as students graduated, so he had to find a way of keeping documentation up to date to enable newcomers and permanent staff members to understand the software at any given moment. This was no simple task, because, by the end of the 1970s, the ISDOS software package contained more than ninety different programs and libraries.

"Since its original version, numerous individuals have directly contributed to PSA development (including faculty, staff, and graduate and undergraduate students), with extensive turnover due to the academic environment, constant retraining of new personnel, and varying degrees of expertise and experience," Chikofsky noted in a 1983 report on the project. "Several versions of the system must be supported simultaneously, matching the software in user organizations. In such an environment, the ability to readily determine the impact of changes is critical to successful management."

Chikofsky now recalls, "So we had to keep track of the software to understand what we had and to be able to use it." He solved the problem ingeniously by using the PSL/PSA system to monitor itself. The same computer-based techniques that had helped programmers write better-specified, better-structured software with fewer errors were also useful for analyzing existing computer programs. "It was the first use of CASE in reverse engineering," he claims.

Chikofsky was not the only one to see the advantages of using PSL/PSA for program maintenance. In 1983, Marylinda Johnson of IBM Information Programming Services reported that IBM had used the same tools to fix an important, widely used software product. As she described it, the program consisted of more than a quarter of a million lines of code originally written in the 1960s in a primitive, difficult-to-decipher computer language. Documentation had failed to keep up with a steady stream of changes as programmers added new features and capabilities to meet user needs. Different groups in different locations had taken charge of the program at various times. Each time something had to be fixed or modified, the programmers working on the task had to spend inordinate amounts of time trying to figure out how this hodgepodge of software worked.

Faced with the pressure to add program enhancements and cut maintenance costs to keep the product on the market, IBM staff turned to PSL/PSA to build up a picture of how the software worked. Using PSL/PSA to analyze the software proved far easier than consulting the code itself to determine how data were used and how information flowed through the system. By the end of this process, the programmers could readily visualize the program's structure—how one piece of code related to another—and determine what particular units of code did. They could identify obsolete pieces that no longer had any function but which no one had previously dared remove for fear of causing unintended repercussions elsewhere in the software.

It was still a major project requiring a huge amount of human effort, but the use of PSL/PSA tools turned what initially looked like an insurmountable task into a feasible endeavor. And it worked.

"An unexpected by-product was improved communication among programmers," Johnson noted. "Everyone could obtain information in personally meaningful forms and know whom to contact for additional explanation."

Despite IBM's success (for a product that the company never identified in public) and Chikofsky's own efforts in publicizing the value of tools such as PSL/PSA for analyzing existing software, few information system managers showed interest. "I spent more than a decade crying in the wilderness, telling people that they could use this stuff for the [software] maintenance effort," Chikofsky says.

The steady increase in the size and complexity of software to scales much larger than one person or a small team can handle had brought simple reading and analysis of programs into disfavor in the 1970s. For a long time, it was just too cumbersome and too labor intensive to perform manual analyses on large projects, and few software technicians felt a compelling need to revisit old software other than for a quick repair or modification.

At the same time, computer-aided software engineering was enduring the growing pains of a new field, as a variety of companies began developing computer programs to help programmers do their jobs. CASE tools were still too new and untested for most programmers to use them with confidence for understanding software, and reverse software engineering was hard to undertake without automated assistance.

Moreover, not everyone agreed that more or improved CASE tools were the answer to the woes of software engineering. It was quite possible, critics argued, that many tools are created to solve problems that arise only because the procedure they are meant to aid is itself wrongheaded. Imagine, for example, that we were still using Roman numerals. Some clever mind would no doubt come up with a computer program to help us deal with this awkward notation. Everyone would agree that using the new tool to simplify a tedious task is just what we need to improve productivity and reduce the probability of error. The much better solution—to replace Roman numerals with a more efficient notation—is overlooked.

Much of computer-aided software engineering may simply entrench bad habits and bad methodologies by making them easier to use. The resulting tools often encourage laziness and discourage true innovation, critics contend. The rapidly increasing speed of microprocessors and the tremendous, continuing growth of raw computer power contribute their share to this trend. Programmers can get away with creating unwieldy, excessively complicated, memory-consuming software because, within months, the improved hardware will compensate for these deficiencies.

By the end of the 1980s, the sheer number of companies trying to cope with legacy systems had grown to the point that the demand for solutions to documentation and maintenance problems had escalated from a whisper to a roar. There was a growing realization that upgrading poorly understood old systems was a central rather than a peripheral concern of software engineering. Small groups of computer personnel in industry, government, and academia began developing tools for automatically extracting useful information from existing, otherwise unintelligible code. They often worked independently and had little contact with others involved in similar projects. The result was a tremendous amount of duplicated work and a cacophony of invented terms to describe scattered, ad hoc efforts.

Chikofsky felt vindicated by the burgeoning interest in reverse software engineering, but he also saw the need for greater coordination and communication among those working on the problem. Even sorting through the Babel of terms that various groups used to describe their work—reverse engineering, design recovery, reengineering, code restructuring, redocumentation, redevelopment engineering, program understanding or comprehension, software visualization—required considerable care and attention. Inconsistencies in the way the terminology was used had become one more daunting obstacle to rapid progress in the field.

The ISDOS project itself proved commercially successful enough to warrant the formation of a company separate from the university to handle the product. Chikofsky became a partner in the new firm and stayed with the company until the end of 1985. He then joined a competitor, Index Technology Corporation, based in Cambridge, Massachusetts, where he served as manager of a research and development group charged with anticipating long-term trends in the nascent CASE industry. As part of his job, Chikofsky helped organize a number of technical workshops so that CASE specialists from around the world could start meeting one another, reporting on their accomplishments, and working together.

Chikofsky took a leading role in organizing meetings and workshops to facilitate the new field's growth. While heavily involved with the Computer Society division of the Institute of Electrical and Electronics Engineers, he wrote articles for society publications and helped keep both computer-aided software engineering and reverse engineering in the limelight. One result was a forum that brought together practitioners in the commercial sector specifically interested in the practical side of applying reverse engineering to software.

He also saw the need for a research component, partly to get improved tools into the hands of the practitioners and partly to evaluate existing tools for their applicability to different types of situations. This was already a serious problem in CASE, where, by the early 1990s, users could choose among hundreds of commercial, specialized, and experimental tools. Bombarded by conflicting, unsubstantiated claims for the capabilities of various products, they had great difficulty selecting the appropriate software wrenches for the job at hand. Many software tools were hard to use; most suffered the same kinds of bugs that plague software in general. It was also hard to tell which tools worked together effectively.

AS CHIKOFSKY BEGAN TALKING to people to organize a reverse-engineering workshop to focus on research issues, he discovered he was not alone. Richard C. Waters, a senior research scientist at the Mitsubishi Electric Research Laboratories in Cambridge, Massachusetts, had become interested in organizing the same kind of meeting. But Waters brought a different perspective to the whole issue of reverse engineering and program understanding.

At roughly the same time that Chikofsky was working on the ISDOS project at Michigan, Waters was completing his dissertation at MIT's Artificial Intelligence Laboratory. His prime interest at MIT was in creating a computer program that could act intelligently as a programmer's junior partner and assistant to help relieve some of the drudgery of a programmer's work. This was the beginning of the ambitious Programmer's Apprentice project at MIT, and Waters became a principal research scientist at the AI Lab to carry it forward. His chief partner in this endeavor was

Charles Rich, who had arrived at MIT in 1973 and had also joined the Artificial Intelligence Laboratory.

Their goal was to develop a theory of how expert programmers analyze, synthesize, modify, explain, specify, verify, and document programs, taking in the full panoply of activities in a programmer's life. From the standpoint of artificial intelligence research, they could use this setting to study fundamental issues of how to represent knowledge and understand reasoning. From the standpoint of software engineering, they were seeking to automate the programming process—an implicit aim of computer science from the time the first programmers faced the vicissitudes of writing code.

For the first decade of their project, Rich and Waters focused mainly on automatic programming as applied to constructing new programs. The idea was to make programming so easy and hence, so inexpensive, that it would generally prove simpler to write new programs to replace obsolete ones, rather than repair them. "That's where the real excitement was for researchers," Waters says. There was little interest in using the tools they developed for analyzing existing software.

By the mid 1980s, however, the immense problems of achieving automatic programming had become clear enough to reveal how distant the goal was. In a thought-provoking 1988 report outlining the prospects of automatic programming, Waters and Rich outlined what they called the "cocktail party" description of the dream: "There will be no more programming. The end user, who needs to know only about the application domain, will write a brief requirement for what is wanted. The automatic programming system, which needs to know only about programming, will produce an efficient program satisfying the requirement. Automatic programming systems will have three key features: They will be end-user oriented, communicating directly with end users; they will be general purpose, working as well in one domain as in another; and they will be fully automatic, requiring no human assistance."

Point by point, Waters and Rich demolished the myths underlying this rosy vision. For example, they attacked the notion that it could ever be possible for an automated system to construct a program for a specific purpose without having complete specifications or a detailed knowledge of the application. It was hard to imagine how any computer could have sufficient knowledge to read a user's mind, in effect, and create a functioning system that does precisely what the user intends, without extensive, detailed communication between the user and programming system.

Nonetheless, they could foresee taking small, modest steps toward facilitating the creation of software. "The automatic programming systems of the future will be more like vacuum cleaners than like self-cleaning ovens," Waters and Rich wrote. "With a self-cleaning oven, all you have to do is decide that you want the oven cleaned and push a button. With vacuum cleaners, your productivity is greatly enhanced, but you still have a lot of work to do."

Thus, programming as a professional activity would not disappear, but the level at which programs are constructed could change, just as it had when programmers went successively from writing codes in machine language, to assembler languages, then to high-level languages such as FORTRAN, and now to programming environments in which they can build software out of code packaged into "objects" that encompass certain characteristics and behave in well-defined ways.

As it became increasingly clear that cheap programming was not just over the horizon, Waters and others began to take a serious look at what the tools they had developed could do in understanding existing software—one of the key aspects of reverse software engineering. Waters himself began to see the need for a meeting to bring together researchers interested in reverse engineering. Under the auspices of the Association for Computing Machinery, he began to put one together. Inevitably, six months into his endeavor, he met Chikofsky, who had a similar project in mind, though under the umbrella of the IEEE Computer Society.

It proved a fruitful meeting of minds. Waters represented the reverse-engineering camp whose primary interest was in the intellectual aspects of software comprehension. Chikofsky represented the practical side, largely oriented toward the development and use of specific tools to solve real-world software problems. Together, they organized the first research-oriented meeting to focus solely on the theory and technology of recovering information from existing software and systems.

In a joint position paper written for the meeting, Waters, Chikofsky, and P. G. Selfridge candidly noted, "In the best of all worlds, we researchers on reverse engineering would be working together toward clear goals of great economic importance. Unfortunately, it appears that we are mostly just groping around in a swamp, each looking for a bit of dry ground (whether or not it actually leads out of the swamp), and running into each other only occasionally."

THE 1993 reverse-engineering workshop concentrated on the nitty-gritty of automated methods of analyzing software. Like the array of wrenches and other specialized tools that a mechanic has on hand to repair a car's engine, reverse engineers usually work with a suite of software tools to glean different kinds of information from a program. Depending on the product, such tools extract various pieces of information concerning how the software under examination is organized, how its different parts may be linked or related, and so on. Often, these characteristics are displayed as diagrams, tables, or charts. With such information, a user can begin to build a mental model of how the software functions.

The meeting participants listed a variety of needs, ranging from surveys to identify successes in the application of various techniques to improved testing of reverse-engineering tools. Just as with CASE tools in general, it's difficult to evaluate how good various tools are, many participants complained. Moreover, because each tool generally has a specific task, software engineers often need a variety of tools to analyze large programs. Unfortunately, many of these tools don't work together very well.

"We're in a field where none of us are experts, but we're learning a lot," Chikofsky remarked after the workshop was over. "The workshop worked very well in bringing together people who had been doing what they didn't realize was related work in different areas."

Someday, it may be possible to tell a bank's computer program about the changes in the tax law, and the program would make the necessary modifications to itself. But that remains a distant goal. "If we started in Boston, and this goal is San Francisco, the tools that we have now have only reached Albany," Waters says. "Exactly how to get farther than Albany is not completely clear."

Most of the present effort has to go into helping the people responsible for maintaining existing software. "In the short term, you have to help them fight the fires that they've got now," Waters declares. That effort by itself could take the next twenty years or longer.

Missing or inadequate documentation, haphazard fixes, widespread distribution, and burgeoning software complexity add up to major headaches for those charged with keeping a computer system functioning. "A lot of us are dealing with leaning towers in our software," Chikofsky says. "We don't really know what the software is."

As an example of how essential reverse engineering can be, one attendee noted that in the early 1980s, information systems personnel in one of the new divisions created by the breakup of AT&T found themselves in possession of thirty business programs written in COBOL. There was no documentation showing what the programs did or how they were designed. They had only the curt, obscure comments that appeared in the code itself. To decide whether to salvage or abandon the programs, software engineers had no choice but to go through the code line by line to extract manually a picture of how the code worked. It was a long, tedious process for the computer experts involved, just to see what they had on their hands.

In reverse engineering, the programmer becomes part historian, part detective, and part clairvoyant. Deciphering an antiquated program may mean learning all over again an obsolete computer language that once flourished and now languishes in obscurity. It could mean trying to read the mind of the long-departed programmer who originally designed the code. It could mean wandering through a maze of plumbing, with redundant branches, dead ends, and hard-to-locate leaks. It's archaeology directed at some of the most complex products ever created by human logic.

Of course, there is a great danger in collecting too much information and spending more time and effort on reverse engineering than the software is worth. "It's very, very easy to get lost in the plumbing," Chikofsky notes.

Nonetheless, "reverse engineering opens up new pathways for us," he says. "You often don't have to throw away the whole system."

In fact, the concept of reuse plays a crucial role in older, more established forms of engineering, whether for building bridges or toasters. More often than not, engineers build what they have built before. "A lot of the things that we prize in other parts of engineering, we don't know how to prize in software engineering," Chikofsky says. "Reuse is one of those principles."

MEANWHILE, the clock continues to run, inexorably ticking toward the first day of the year 2000. Practically everyone in the computer community knows about the potential time bomb lurking in antiquated software. But most are too busy with the more immediate concerns of keeping systems going, troubleshooting glitches, improving performance, and finding additional storage space for proliferating data to heed the ticking clock.

When they do think about the impending date problem, system maintainers see a huge, unappealing job ahead. Given the complexity of date usage in software, not even CASE tools provide a quick, complete, satisfactory answer. Moreover, no one has a reliable estimate of how pervasive and disruptive the problem will be. An internal survey at one large, high-tech company revealed in 1992 that eighteen out of the organization's 104 computer systems would fail in the year 2000. The eighteen systems, all vital to the company, consisted of 8,174 individual programs and data-entry screens and 3,313 databases. Correcting the deficiencies would likely require far more time than the ten weeks it took just to identify them. Many other businesses and agencies face similar situations.

In 1992, the failure of Tandem CLX computers because of a clock problem provided a hint of what may occur on a vastly larger scale at the beginning of the year 2000. The Tandem computers are specifically designed to be fault tolerant. They incorporate backup circuitry and software that allow the system to keep on working despite the failure of various components. At the very least, they capture the data at the moment of failure so the system suffers no significant information loss in the event of an abrupt, unavoidable shutdown. These qualities make Tandem computers a popular choice for automated teller machines and for electronic funds transfer terminals, which allow retail customers to pay for purchases directly out of their bank accounts instead of writing checks.

On November 1, 1992, at precisely 3 P.M., all Tandem CLX computers in New Zealand suddenly stopped functioning. Luckily, it was a Sunday, and business was light at the gas stations and supermarkets most affected by the computer failures. While computer experts in New Zealand were struggling to identify and correct the problem, it hit Australia, then Japan and other parts of Asia, eventually sweeping around the entire world in step with the shifts from one time zone to the next. This pattern suggested that the glitch was somehow triggered by the particular combination of numbers registered in the computer when its internal clock reached the fateful moment. A bug in the software caused this string of numbers to be interpreted wrongly.

By the time the critical hour arrived in Europe, Tandem engineers had worked out temporary solutions. Users could roll the computer clock forward past 3 P.M., then shift back to the correct time later. Or they could wait until 3 P.M. arrived, and a moment later, restart their systems from scratch. Users prayed that this cavalier treatment of time would cause no disruptions elsewhere in the software or in the record keeping.

The 1992 incident wasn't the only date problem to disrupt Tandem computers. On August 27, 1991, at 4:22 P.M., a clock-related flaw in a newly revised version of the Tandem Safeguard security system software put Tandem VLX and Cyclone computers out of commission. A faulty section of code in the security system interpreted a unique combination of numbers corresponding to this particular date and time as a command that it couldn't implement. The corrupt logic forced the computer's manipulations into an endless loop, effectively tying it up. The security system then locked up the computer, permitting no one to log on—not even those who were trying to fix the problem. The failures occurred first in New Zealand, then in Asia and Europe, before Tandem staff identified the problem and started notifying users that they could avert a shutdown by turning off the security system between 4:22 P.M. on August 27 and 5 P.M. on August 28.

Clock problems are by no means limited to Tandem computers. Peter Neumann's Risks forum provides a steady diet of calendar, date, and time malfunctions, ranging from months with the wrong number of days to faulty corrections for daylight savings time. But nothing in these annals suggests the magnitude of the potential disasters lying in wait for computer users on January 1, 2000.

In a 1990 editorial titled "T minus 10 and counting," Chikofsky wrote: "What can we do to prevent disruption? Well, besides declaring January 1 and 2, February 29, and March 1 in the year 2000 to be international business holidays for clock resetting and database reprogramming, we can begin now to conduct a comprehensive audit of date usage in our existing systems and new designs." He concluded, "We'd better get started. This may take the full ten years."

The effort has barely begun. Pessimists point to the potentially disastrous consequences of delaying action. Optimists note that software is constantly being updated and replaced, usually quite painlessly. In some instances, companies are even opting to rewrite their software from scratch. They get rid of unwieldy, antiquated code, fix the date problem, and incorporate a host of other enhancements, all in one fell swoop.

The date problem and the travails of fixing it have also attracted the attention of science fiction author Arthur C. Clarke. In his 1990 science-fiction novel *The Ghost of the Grand Banks*, he described his solution: "There were not enough programmers in the world to check all the billions of financial statements that existed, and to add the magic '19' prefix wherever necessary. The only solution was to design special software that could perform the task, by being injected—like a benign virus—into all the programs involved."

Of course, all did not go smoothly. Clarke continues: "During the closing years of the century, most of the world's star-class programmers were engaged in the race to develop a 'Vaccine '99'; it had become a kind of Holy Grail. Several faulty versions were issued as early as 1997—and wiped out any purchasers who hastened to test them before making adequate backups. The lawyers did very well out of the ensuing suits and countersuits."

For those programmers getting a little nervous about the arrival of the year 2000, one wag has suggested a programmers' cruise, which would depart on December 30, 1999 for a thirty-day excursion. To ensure peace of mind, the cruise would come with certain guarantees. The ship would be operated by mechanical or simple electrical controls, with no computers. The crew members would be selected on the basis of their expertise in dead reckoning and celestial navigation. The ship would avoid sailing under established airline routes and would stay away from normal shipping lanes. There would be no communication with shore.

Because many people erroneously believe that 2000 will not be a leap year, it may be advisable to extend the programmers' cruise into March.

Whether or not the cruise ship sets sail, the date problem remains an embarrassing, uncomfortably prickly, hard-to-remove thorn in the paw of software engineering. No matter what happens, reverse engineering will have to be part of the solution.

Moreover, even without the date problem, there's no longer any shortage of demand to scrutinize obsolete, barely understood software dinosaurs. "Reverse engineering is here to stay," Chikofsky insists. "It is not an option for us in the software engineering community."

# CHAPTER 6

## Sorry, Wrong Number

**CONTRASTING SHARPLY** with its stately neighbor along the fringe of Mineral Circle on the Berkeley campus of the University of California, the spartan façade of Evans Hall suggests utility, economy, and modernity. Its drab, barren concrete is overshadowed by the ornate, time-tested splendor of the Hearst Memorial Mining Building, whose mass of stone, brick, and tile inevitably draws the eye.

But Evans Hall has an insistent appeal of its own. Stepping out of the elevator at the fifth floor, a visitor sees an enormous wooden frame mounted on the wall opposite the elevator doors. Wooden beads the size of hefty cantaloupes appear poised to slide along thick metal rods that cross from one side of the frame to the other. A fitting tribute to computation, this giant abacus serves as a totem for the computer scientists whose offices line the corridors.

The hallway walls sport a motley collection of the mundane and the revolutionary, as befits a center for education and research. Haphazardly posted scraps of paper list student grades; suggest solutions to assigned problems; and leave instructions on office hours, classroom changes, and other minutiae of the academic life. At the same time, paper sheets large enough to cover a desk commemorate in their intricate lattices of thin lines the design of innovative microprocessors and novel computer architectures.

Rows of faculty portraits arrayed on the wall near the main office hint at the human side of this endeavor. But there's a curious anomaly here. Instead of the usual photograph, the place reserved for William M. Kahan features a drawing of a wolf. Ever camera shy, Kahan allows only this reference to his childhood nickname of "Velvel" (meaning "little wolf") to represent his membership in an illustrious faculty.

On the morning of October 16, 1992, Kahan was standing outside his office. It was the day before a conference and banquet that had been organized to celebrate his impending sixtieth birthday, an honor he was sharing with Beresford N. Parlett, another Berkeley faculty member and esteemed veteran of the computer age's adolescence. Not uncommon in the academic world, such celebrations represent an opportunity for former students to pay tribute to their mentors and to show off their own accomplishments in extending and disseminating the lessons of their lengthy apprenticeships as graduate students.

Kahan's large, casually accoutered figure loomed over a student who was standing before him seeking advice. Boxes overflowing with disheveled papers lay in a heap near his feet. To a passerby who glanced quizzically at the disorder, Kahan wryly noted that his office had suffered a minor earthquake—a local shudder that had toppled parts of the towering stacks of papers, reports, documents, and books that filled nearly every available space in his narrow work area. A glimpse into the room through the open door served only to confirm the impression of a man-made cave, made claustrophobic by enormous paper stalagmites.

A decade earlier, there had been more breathing room. Kahan had been able to welcome a visitor into his office to sit and ponder the vagaries of calculation. Even then, he had a formidable reputation for a legendary exactitude born of a lifelong passion for clear thinking. And the victim of his scrutiny at that time was the humble pocket calculator.

Ever ready to perplex, astonish, amuse, and perhaps alarm an attentive visitor, Kahan kept a desk drawer well stocked with miscreant calculators. Look at this wonder, he remarked with playful irony, as he reached in and pulled out one of these models. Nimbly manipulating the tiny keys, he teased from the calculator a sequence of numbers that ended by demonstrating—without a slip of the finger—a flaw or some weird quirk in the way it performed.

For instance, different financial calculators manufactured by reputable firms can produce conflicting answers in what appear to be relatively straightforward situations. Consider the calculation of the future value of funds deposited daily and accruing interest at a certain rate, say 365 payments of $10,000 each at an interest rate of 3.6500364 percent per annum, compounded daily.

For these perfectly reasonable values, one calculator gives the answer $3,717,213.88, another displays $3,717,204.03, a third shows $3,717,241.83. In the first two cases, the calculators end up "embezzling" $27.93 and $37.78, respectively. The third calculator's answer happens to be only 2 cents away from the true answer, correctly rounded to the nearest cent.

Though such discrepancies represent just a small fraction of the total sum involved, they are large enough to attract the attention of examiners at banks, mortgage companies, and other institutions, potentially leading to investigations on suspicion of financial fraud. They are also bothersome enough to undermine confidence that no bigger discrepancies lurk in other types of calculations performed on these calculators.

To a stockbroker or businessman, using such a calculator means little more than keying in the necessary data, then pressing the "solve" key or equal sign. Few realize the subtlety of the mathematics that goes on behind the calculator's plastic façade in the tiny fractions of a second before the display lights up with the answer.

"Should the businessman trust all the figures displayed by his business calculator?" Kahan muses. "If not all, then which? Must he take a college course to find out?"

It may seem strange to think that computers, from pocket-size calculators to room-filling number crunchers, would have trouble with arithmetic. But calculators and computers occasionally give completely erroneous results for innocent-looking problems, and sometimes the false answers appear quite plausible. Indeed, funny things can happen with even the most sophisticated products available, although they manage to compute a reasonable result in nearly every other case.

This curious situation stems from the fact that calculators and computers are designed to manipulate numbers having only a limited, fixed number of digits. Each of the numbers of a calculation has to fit into a certain number of slots, and how many slots are available to store a number may vary from computer to computer. A calculator display, for example, has spaces for precisely eight, ten, or some other number of digits—and no more—though the logic circuitry inside the calculator may also keep track of a few extra digits that aren't displayed. This intrinsic limitation of the way in which computers handle numbers has several important consequences.

What's the largest whole number you can think of? There really is no such thing because you can always use ordinary, everyday arithmetic to generate an even larger integer than the one you chose initially. Simply add 1 or multiply by 2.

But a computer or calculator must have a largest possible number. No whole number can have more digits than will fit into the definite number of spaces allocated for each number stored and processed in a given computer. This number may be huge, but it nonetheless represents a strict upper limit. Multiplying this number by 2 gives an answer with more digits than fit in its designated space. It results in an "overflow."

Similarly, every computer has a smallest possible number. Dividing this number by 2 produces a fraction, or decimal, that isn't zero, yet the computer can't store the new number. The result is an "underflow." Unable to represent the real answer, the computer may simply call it zero.

Computer users always have to be careful in making sure that any numbers that come up during a chain of calculations don't stray outside the bounds set by the largest and smallest numbers that a given computer can represent. Their programs have to be written in such a way that numbers stay within a specified range no matter what numerical contortions are involved—often a rather tedious, but essential, bookkeeping chore.

In the past, when calculations involved decimal fractions, users generally had to keep track of the decimal point's position as a separate operation. For example, multiplication of 2.36 by 45.81 required the use of whole numbers ($236 \times 4581$) in the computer, and the computer would generate the answer 1081116. The user could then put the decimal point in the right place to get the actual answer of 108.1116.

But the use of whole numbers (or integers) proved too limiting for many applications. In the 1950s, to extend the range of numbers that a computer could handle, mathematicians and computer scientists began turning to a "floating-point" number system. The idea was to represent any given number as a decimal fraction followed by a second set of figures indicating by how many tens the decimal fraction must be multiplied to give the actual number.

For example, 2,380 is equivalent to $0.238 \times 10 \times 10 \times 10 \times 10$ (or $0.238 \times 10^4$). You can represent this number on the computer as the pair 00000238 followed by 04, and it will still fit snugly in its set of designated slots. Similarly, 23,800,000 (or $0.238 \times 10^8$) would show up as 00000238 and 08.

This numerical subterfuge greatly expanded the range of numbers available to computer users. But it did not eliminate the problems of overflow and underflow during calculations. It was still possible to obtain numbers that exceeded the bounds allowed by this more generous representation.

And there were some new problems. Computer programmers had to learn how to round off numbers properly to keep calculated answers on target. For example, there's no exact way to write the fraction one-third ($\frac{1}{3}$) as a finite decimal fraction. No matter how many threes are added to the end of $0.33333\ldots$, this expression is never precisely equal to one-third. Because both computers and calculators typically work with numbers having a definite number of digits, decimals consisting of an infinite or even lengthy string of digits must be unceremoniously chopped off, with or without rounding, after a defined number of digits to fit the machine's format.

This chopping leads to some interesting peculiarities. For example, in this "approximate" arithmetic, the order in which numbers are multiplied can sometimes make a difference to the answer, contrary to the normal rules of arithmetic, in which $3 \times 2$ is always the same as $2 \times 3$.

In the 1950s, when floating-point arithmetic was first introduced to computation, many people were disturbed by such anomalies. The complexity and subtleties of the bookkeeping required to avoid these situations were sufficiently mysterious that Alston Householder, one of the pioneers of numerical linear algebra and a prominent figure in the applied mathematics community, once publicly declared that he would never fly in an aircraft that had been designed with the help of floating-point arithmetic. The accuracy of such computations could not be trusted, he insisted.

But this approximate, floating-point arithmetic, despite its apparent shortcomings, does work amazingly well. The reason, says Kahan, is that the approximations occur in predictable ways. In spite of the fact that many mathematical rules of ordinary arithmetic are broken, certain kinds of rules still apply. Hence, approximate computations can produce correct answers. It can do so even in cases in which the intermediate answers in a lengthy calculation may appear completely wrong.

"Let's suppose you have a big computation," Kahan says. "As

the machine is running, you hit the stop button. You record all the numbers in the machine, and then you let the calculation go on to the end."

It's quite possible, he continues, that when you stopped the machine and recorded all those numbers, there wasn't a single number that agreed in any digit with the numbers you would have gotten if you hadn't rounded them off as you went along. Yet the final result is correct to within the last few decimal places.

"In other words, even when the computation appears to have gone wrong, it seems to come out right in the end," Kahan states. "This miracle is the rule rather than the exception." Most modern computation would be impossible without it.

To Kahan, floating-point computation is like a delicate web, and the necessary approximations (or roundings) correspond to the cutting of certain strands of the web. The web as a whole holds together even though from time to time a few strands need to be cut.

"If you were told not to break any strands, to do everything exactly, you would have to wait to the end of the universe for your result," Kahan says. "So we've got to make approximations, billions and billions of them, to get anything done."

But once in while, the wrong strand gets cut, and the web collapses. The final numerical answer is seriously wrong. "The trouble is that it isn't easy to say which strands you can break and which ones you can't," Kahan admits.

Discrepancies in calculations arise in a variety of ways. Try the following experiment on a rudimentary, four-function calculator. Key in 1, then a division sign, then 11, followed by another division sign. The calculator display should show something like 0.090909. Now press 100 and a division sign. The display gives 0.000909. Divide by 100 again, and the calculator shows 0.000009. Each time, additional zeroes appear on the left-hand end of the digit string and several digits disappear from the right-hand end.

Now multiply this number by 100. A superior calculator regenerates the answer 0.000909, but many models produce the number 0.0009, which means that the designers of these particular products chose simply to truncate decimals having lots of digits rather than preserve internally the extra digits that don't fit on the display. Multiply the answer by 100 again, and the results may get even more erroneous.

On a more sophisticated calculator, try taking the square root of 2, then the square root of the answer, and so on, for a total of, say, ten or even a hundred times. When it's done, square the answer (multiply it by itself) the same number of times, and check how close the final answer is to 2. The discrepancy is sometimes startling.

"What you see is often not what you've got; what you've got is often not what you want; and what you want is often not what would be good for you," Kahan declares. "That's just a little bit disconcerting."

OF COURSE, not all of the erroneous numerical results that computers and calculators may disgorge result from the workings of floating-point arithmetic. Sometimes, humans simply make mistakes.

In 1982, several purchasers of the widely heralded IBM personal computer, which had just come on the market, were surprised to find a particularly blatant error. The computer said that 0.1 divided by 10 equals 0.001 instead of 0.01.

"It's the type of thing where you sit and look at the terminal, and your mouth just drops open," said David Walonick, the computer programmer and consultant in Minneapolis who discovered the error. "I was running test data through my package, and it was coming up with wrong answers."

At first Walonick had difficulty in persuading IBM that the computer was making a mistake. "When I called them, I was told that beginning programmers have problems like that," Walonick said. "They weren't even willing to try it when I first called."

That's the kind of unhelpful response more than a few perplexed and stymied computer users have encountered. Indeed, making the customer feel guilty has a venerable history in the computer industry, and many products have survived buggy births because purchasers often ended up blaming themselves for the problems they encountered.

A story in *The New York Times* about Walonick's discovery brought a response from IBM and an admission that its computer was actually at fault. To circumvent the problem, IBM released a corrected version of the computer's operating-system program.

The mistake arose from an error in the logic circuitry built into the computer's microprocessor chip. Most computers (and calculators) operate in the binary number system, in which all numbers are represented by strings of ones and zeroes. In the binary system, 10 stands for 2, 11 for 3, 100 for 4, 101 for 5, and so on. When a computer or calculator is instructed to perform an arithmetical operation such as 2 + 3, it converts the numbers into binary form, performs the addition, then translates the result back into decimal notation for the benefit of users.

In the case of the IBM personal computer, a designer made a mistake in the conversion of binary to decimal notation in one special case. The computer calculated accurately in binary, but it displayed the wrong decimal answer. The misplaced decimal point appeared only in isolated instances under unusual, narrowly defined conditions that arose when programmers were using a part of the machine's software that converts a program written in the computer language BASIC into strings of instructions that the computer's circuitry can understand.

Subsequent microprocessors also suffered arithmetical glitches. In 1994, users of desktop computers based on the Pentium microprocessor discovered that the chip gave incorrect answers to certain floating-point division problems. For example, dividing 5,505,001 by 294,911 produced the answer 18.66600093 instead of 18.66665197.

It turned out that only 1,738 numbers out of 64 trillion combinations of numbers actually triggered the bug. So the chance of randomly coming across these particular numbers was small. But such errors could matter in scientific and engineering applications and even in some spreadsheet calculations.

The chip fault itself had occurred because of an omission in the translation of one mathematical formula into circuitry on the chip. This omission meant that, in certain cases, the flawed Pentium chip rounded numbers off incorrectly, saving only 5 instead of 16 digits.

Some arithmetical peculiarities are simply the result of deliberate compromises made by computer programmers. The calculator built into the popular and widely distributed Microsoft Windows program displays a number of quirks. For example, in calculating 750.35 − 750.30, it produces 0.04999999999995 instead of 0.05 as the answer. To Kahan, this behavior merely exemplifies his maxim that "what you see ain't what you've got."

Instead of working in decimal digits, the software underlying the Windows calculator converts numbers into their floating-point binary equivalents, but it can't make the conversion exactly. So the decimal-to-binary conversion changes the numbers slightly. Thus, when the program computes the difference, it gets something a little different than what you would expect (but you can't see it because it all happens inside the computer). The program then converts the result back into decimal notation with reasonable accuracy.

There's nothing really wrong with the calculator. It does its job quite well. But programmers at Microsoft might have forestalled the development of a minor niche industry supplying replacement calculators to those bothered by such shenanigans by basing their own program on decimal instead of binary arithmetic to avoid conversion problems.

Casio calculators show another peculiar effect. If you divide 1 by 3, then multiply the answer by 3 and subtract 1, you get 0. But if you divide 100 by 3, then subtract 33, then multiply by 3 and subtract 1, you don't get 0. Depending on the calculator, the answer that appears is something like −.000000001.

What happened is that the people who designed the Casio calculators used a finite string of threes after the decimal point to represent ¹/₃. Multiplying this number by 3 gives a string of nines as the answer. But seeing such a result would probably upset customers, so the answer gets automatically changed to the "correct" result. Kahan calls this "cosmetic" rounding because it attempts to hide blemished arithmetic. The blemishes surface when a slightly more complicated calculation, for which the designers did not make provision, is done.

Similarly, Kahan can demonstrate that on several machines, one-third (¹/₃) is plainly not equal to nine twenty-sevenths (⁹/₂₇)! All of these quirks stem from somewhat arbitrary decisions made by calculator designers about the number of digits that a particular calculator must keep track of during intermediate steps in a calculation and the casual use of cosmetic rounding.

But do these tiny errors, these quaint quirks really matter? "To any sensible man, the answer has to be 'no, they can't possibly matter,' " Kahan insists with a devilish twinkle. "But there are examples where they matter an awful lot."

By handling numbers in unexpected ways, however rarely, computers and calculators create headaches for computer programmers and set traps for unsuspecting users. The results of this perverse arithmetic may include mangled calculations of bond yields or flawed estimates of the strength of a beam in a building. And when computers operate heavy machinery, calculation errors may result in more than defective numbers. Incorrect results channeled to a robot could prompt a wild swing, dangerous to both product and machinery.

One well-known optics company spent nine months and a lot of money looking for mechanical problems in an apparently faulty automated measuring instrument, only to discover that the fault lay in the inability of the machine's electronic controller to compute accurate values of angles.

Numerical glitches have also affected spacecraft maneuvers. In May 1992, the crew of the orbiting space shuttle *Endeavour* tried to rescue an Intelsat communications satellite, an operation that required it to approach and grab the satellite using a spindly robot arm. As on two previous maneuvers to rendezvous with the errant satellite, the astronauts fed radar and star-tracking data into the shuttle computers to calculate the trajectory, fuel, and rocket requirements for the final approach. But this time, the software failed to compute the needed data. Instead, the computer kept signaling that the number of steps it required to reach an answer exceeded preset limits—limits designed mainly to keep the computer from being tied up solving a single problem. Because it wasn't immediately apparent why the computer couldn't do the necessary calculation, this unexpected glitch threatened not only the rendezvous but also the entire shuttle guidance and control software.

"Even though we knew from the targeting standpoint we could provide the targeting solution from the ground, we were concerned about the safety of all the other software on the orbiter—if the computers don't work, nothing on the orbiter works," said Wayne Hale, one of the mission directors quoted in an article on the incident in the May 25, 1992 issue of *Aviation Week & Space Technology*.

A combined team from IBM, NASA, and Rockwell International mounted a frantic, intensive effort to determine whether the problem affected all of the software or was restricted to just one of the forty-seven different subunits in the rendezvous program. Within an hour, they discovered that only a small part of the software was affected, and they decided the mission could proceed as planned. The astronauts managed to snag the satellite.

Engineers eventually traced the software problem to the sensitivity of equations used to do the calculations to certain numerical values that happened to come up at the crucial moment. In this case, the software failed to provide an answer because a different number of digits was used in the values describing the shuttle's position and velocity than in the numerical values establishing whether a calculated answer had finally reached the correct precision. Because these two sets of numbers never precisely matched, the software couldn't conclude its set of calculations.

This curious anomaly arose from a questionable programming practice allowed at IBM but not at Rockwell. Normally, programmers specify how many digits should be used for any value assigned to each variable in an equation. IBM practice permitted a so-called single-precision value (only eight digits long) to be attached to a double-precision variable—one that would normally require a sixteen-digit value. So when IBM programmers made some last-minute software changes before the *Endeavour* flight to facilitate the planned rescue, they used this shortcut. Unfortunately, the Rockwell software components did not recognize this particular subterfuge, and the resulting mismatch between data and software parameters derailed the calculation. A subsequent report detailing the error suggested that better communication among the contractors could prevent similar mishaps in the future.

A tiny numerical error was also involved in the failure of a Patriot antimissile battery on the outskirts of Dhahran in Saudi Arabia to intercept an incoming Scud missile during the Gulf War (see Chapter 1). The Patriot system tracked its targets by measuring the time it takes for radar pulses to bounce back from them. Its timing hardware relied on a digital clock that allocated twenty-four binary bits (ones and zeroes) to register the time. The original programmers decided to record the delays of echoed pulses in tenths of a second. But one-tenth can't be expressed exactly as a binary number. So the conversion from binary to decimal notation (from 0.1 to 0.0001100110011 . . .) introduces a slight truncation, or rounding, error that causes the stored time to drift away from "real" time by 0.3433 second every hundred hours.

Normally, because tracking requires the calculation of time differences rather than direct use of the actual time, this small discrepancy would essentially cancel out. However, the Patriot's original software had been written twenty years earlier for tracking aircraft, and it had been modified several times since so that the weapon system could cope with the much higher speeds and steeper trajectories of ballistic missiles. One of the software modifications introduced a routine for converting the time from binary into decimal form more accurately than before. But the more accurate version wasn't used everywhere in the software. It turned out that in the time-delay calculation, a less accurate, truncated system time was subtracted from a more accurate time of a radar pulse. The truncation errors no longer cancelled, and the discrepancy increased steadily the longer the system was in operation. The Patriot missile ended up missing its target.

THE UNSETTLING UNCERTAINTY built into computer arithmetic creates tough problems for programmers. It's possible, for instance, that if a programmer happens to include a test to check whether an answer (say, the result of multiplying $1/3$ by 3) is really 1, the computer will say no because 1 and the internally stored number 0.99999 . . . 9 are not the same. Yet, suppose that the difference between this answer and 1 (which, according to the test, is not zero) must be divided into another number. The computer may very well regard this difference as zero and refuse to continue the calculation because dividing by zero is against the rules of arithmetic. So the computer first lies to you, and then it punishes you for believing the lie!

It's not that mathematicians and computer scientists have no way of mitigating the effects of round-off and other numerical errors. Indeed, they have developed a variety of methods for guarding against these uncertainties and inaccuracies. The trouble is that they disagree—sometimes passionately—on precisely how this arithmetic should be done under which circumstances and how and when numbers should be rounded off.

This diversity of approach adds to the complexity of computer programs and of integrated-circuit chips designed to facilitate arithmetical operations, often so encumbering otherwise routine calculations that a computer's number-crunching pace slows to a crawl. For many programmers, the situation becomes a trade-off between accuracy and efficiency. And in the absence of universally accepted standards, the engineers who design calculators and write computer programs have carte blanche to do whatever they deem appropriate.

Indeed, a stockbroker or mortgage lender doesn't care whether a business calculator or a computer spreadsheet uses logarithms or some other arcane arithmetical operation to determine a bond yield or a mortgage payment. All that matters is whether the final answer is correct. But on many of today's machines, there are no such absolute guarantees. Unfortunately, users require a relatively high level of mathematical sophistication to ascertain whether they can rely on a calculator's or computer's answer to a particular problem.

"Somehow, we have to get mathematical ideas into packages that can be used safely without obliging users to understand all the details," Kahan says. His approach reflects his broad perspective on the problem. "There *are* mathematical rules, even though they are not altogether obvious, which if you violate them will cause pain," he emphasizes. "And if you don't violate them, you get miracles—things work despite the fact that we know that what we're doing is wrong."

Kahan has built a career out of worrying about minute details that most other people neglect or dismiss because they believe such fine points don't and can't matter. To emphasize his concern, Kahan started off a 1972 paper on error analysis with an oft-cited aphorism, taken from *Poor Richard's Almanac* by Benjamin Franklin:

"A little neglect may breed mischief . . . for want of a nail the shoe was lost; for want of a shoe the horse was lost; and for want of a horse the rider was lost."

Kahan went on, "A horse, a rider, a battle, a crown; that they all might be lost for want of a nail is plausible though unlikely. How likely is anything important to be lost because of a rounding error?"

To illuminate the issue, he introduced a favorite metaphor: "There is a natural analogy between illness and numerical inaccuracy. Germs and rounding errors are small, numerous, and best combated by sanitary precautions which, alas, are all too frequently neglected, not so much because of their intrinsic difficulty or expense as because of indifference or ignorance. When that neglect breeds mischief, the doctor is called."

But as a meddlesome mathematician stepping on professional toes, Kahan recognized that his comments and criticisms—which could sound to others as extraordinarily nitpicky, pedantic, and possibly moralistic—could easily alienate just those people he was most eager to reach. In his own defense, he often cited the stories of the Viennese obstetrician Ignaz Semmelweis and the English surgeon Joseph Lister who, in the 1860s, had suggested that surgeons should wash their hands before operating.

Because surgeons did not understand that the lack of cleanliness contributed to disease and death, this suggestion was poorly received. Philosophically inclined medical men of the day argued that the risk of death was intrinsic in surgery and hence unavoidable. And they were quite correct. Even as they uttered this truth while avoiding the extra costs in time and expense of washing hands, donning a clean coat, and boiling medical instruments, they made the truth more true. The prevalence of dirty arithmetic in computation has had a similarly negative impact.

But to leaven his argument, Kahan notes that the analogy isn't perfect. Germs are actually much more persistent than rounding errors. It is possible to deal with rounding-off in a comparatively routine way that would be the envy of medical practitioners. There *are* cures for faulty arithmetic.

Kahan likes to tell the numerical "horror" story of what happened to a University of Toronto graduate student in the early 1960s. It was a time when students, professors, and researchers were encouraged to take advantage of the computing power available on campus, and many of these novices tended to take for granted the various "canned" library programs available for their use, often ascribing to these numerical programs an unjustified infallibility.

The aeronautical engineering student had come up with an idea for modifying the design of a short takeoff and landing aircraft to prevent the sudden onset of stall, or loss of lift. To test his idea, the student wrote a program for an IBM 7090 computer to simulate the way air flows over a wing. He ran his simulation several times, and each time, the results indicated that his novel wing design would not prevent an aircraft from stalling abruptly at low speeds. He got the same negative result when he instructed the computer to double the precision of all its calculations by using, in effect, twice as many digits for each operation.

Around this time, as a young assistant professor at the university, Kahan was in the middle of tracking down the cause of some inaccuracies he had observed in the way the IBM 7090 computed logarithms. He tinkered with the computer's logarithm recipe, in the end writing his own version of the logarithm program. Displaying a characteristic prudence, he tested his substitute by rerunning several computer programs that had invoked the logarithm operation when going through the computer in the previous week. Only two programs, including the one belonging to the aeronautics student, produced substantially different results. This time, the student's wing design worked.

But why had the presumably more accurate calculations using double-precision arithmetic produced the opposite result? Part of the answer arrived a week later when IBM issued a revised version of the double-precision arithmetic package for the 7090, which corrected several flaws that had become apparent in the original software. Using this improved software, the student ran his simulation again. This time, the results demonstrated that his innovative wing design would keep an aircraft safely aloft at low speeds. Now he had a thesis topic.

But he faced one more scare before he escaped with his degree. As he was putting the finishing touches on his thesis, the university replaced its 7090 with a more advanced model, the IBM 7094. This new machine came equipped with a procedure for doing double-precision arithmetic built into its circuitry. On this computer, the wing design failed again—just as it had many months earlier on the older machine. This disaster precipitated a frantic search for the cause of this puzzling failure.

It turned out that the fault lay in the way the 7094's double-precision hardware dealt with subtraction. Whenever the computer had to subtract 1 from a number ever so slightly smaller than 1, it prematurely discarded the smaller number's last digit (fifty-fourth bit) in order to complete the subtraction. In other words, there was no guard digit—no extra digit stored internally (but not displayed)—to ensure accuracy during computations. In certain cases, the absence of this feature changed the answers substantially. To cure his problem and save his thesis, the student turned to an arithmetical subterfuge. He programmed the computer to subtract 0.5 twice whenever it had to subtract 1 from a number slightly less than 1. Ironically, the revised double-precision software for the 7090 had provided for a guard digit, so this problem hadn't come up on the older computer.

After graduation, the student went to work for de Havilland, an aircraft manufacturer well known for its Dash-7 airplane and other short takeoff and landing craft. When he ran his simulation on the company's Univac 1107 computer, the fledgling aeronautical engineer saw to his horror that the wing design no longer worked. It was months before he discovered that the computer, instructed to "clean up" sloppy software, had automatically transformed 0.5 sub-tracted twice into the more efficient expression of 1 subtracted once, and the original guard-digit problem reappeared.

"How often are engineers baffled by subtly wrong computations, thwarted in otherwise exemplary endeavors, and unable to uncover what went wrong?" Kahan wonders. "And how often is the engineer who expresses doubts about the computing system he must use regarded as if he were Dante's bad blacksmith?"

Kahan has argued, "The real horror of this situation is that the incidence of error in one's final conclusion is unknowable. We do not know how often numerical results are considerably more wrong than is believed by the people who use them."

Such perplexities arise out of decisions that designers of computers and calculators make routinely, constrained by the technology and their own experience and expertise. Most designers are not mathematicians but electrical engineers who take pride in the fact that they can figure out a way to do something a little bit faster than their competition, and they don't seem to worry as much about the mathematical consequences of their ingenuity. They often don't realize their reasoning may be faulty or shortsighted until their ideas are already etched in silicon, and then it's too late.

Cleaner arithmetic than has been customary in big computers and little calculators carries a cost in terms of greater complexity and decreased efficiency of computation, but the attendant benefits make the effort worthwhile, Kahan argues. Serving occasionally as a consultant to Hewlett-Packard's calculator division, he had a chance to influence directly the design of a new line of calculators, and he remains proud to this day of how cleanly those calculators work.

Kahan's anecdote about the aeronautical engineering student also illustrates how even apparently trivial differences in the way different brands and models of computers perform arithmetic and round off numbers cause extraordinary headaches for programmers who want to write software that will be compatible with a variety of machines. Because a program written for one company's computer will not necessarily run successfully and correctly on another's, programmers must learn each computer's idiosyncrasies and program around them by providing tests and alternate routes to make sure that the computer does not stop in its tracks, unable to cope with an unexpected division by zero or some other problem.

"What makes tests and branches expensive is that programmers must decide *in advance* where and what to test," Kahan says. "They must anticipate *every* undesirable condition in order to avoid it, even if that condition cannot arise on any but a few of the machines over which the program is to be portable."

Examples of anomalies abound for every brand of computer.

Early models of the Cray family of supercomputers could readily divide one number into another. But every once in a while when both the divisor and dividend were very small, the computer would signal an overflow, meaning that at some stage, the calculation produced a number larger than the computer could handle. Yet a check of the data would reveal that the final answer should be quite reasonable.

The problem occurred because these computers performed division by first dividing the divisor into 1 (that is, by finding the reciprocal of the original number), then multiplying that result by the dividend. Of course, dividing a very small number into 1 can produce an enormous result—sometimes a number so large that it exceeds the computer's limit, and the calculation cannot proceed.

When IBM first introduced its highly successful 360 line of computers in the 1960s, the company focused most of its efforts on the commercial data processing and business market, and it initially paid relatively little attention to how these machines performed arithmetic. However, users interested in scientific and engineering applications soon discovered major flaws, including the absence of guard digits for high-precision computations and careless rounding off of numbers in certain situations. Indeed, it was sometimes possible that a given number multiplied by 1 would not equal the given number. Eventually, such vexing problems were fixed, but it took a lot of complaints from users and a lot of time for this to be accomplished. Now, the 360/370 family of computers, though superseded in the marketplace by more advanced products, has remarkably clean arithmetic.

The list of numerical anomalies in computer designs goes on and on, covering just about every computer on the market. "Programmers take pride in coding around these perversities," Kahan notes. "Unfortunately, we may not have reckoned the cost."

To Kahan, forcing programmers into such contortions squanders their talents—a precious and rare resource—especially when programs have to be rewritten again and again to accommodate the quirks of each new computer on which the program is to be run. Moreover, in addition to the programmer's wasted time, some programs end up accepting an unexpectedly limited range of data, some are less accurate and more complicated to use than they ought to be, some are less helpful than users would like when things go wrong, and some are simply slow.

"Little tricks here and there add up to complicated programs," Kahan remarks. "The more complicated the program, the more vulnerable it is to blunders and the more it must cost to develop."

IN THE LATE 1970s, the idea of creating a standard to do away with the niggling discrepancies in the way different computers handled arithmetic had been brewing for a number of years. Proponents of such a standard argued that it would encourage the development of more efficient numerical software, stimulating demand for computers. It would permit the mixing and matching of chips fabricated by different companies. And it would minimize argument over whose arithmetic is better. Vendors could concentrate on price and performance instead of waging a futile warfare over arithmetical schemes that weren't terribly good to begin with.

But it was clear from the beginning that such an effort would meet considerable resistance. Manufacturers of computers had a strong vested interest in doing things just the way they had always done them. Designers of computer chips had valid reasons for their choices, and each company wanted to make sure its future models would be compatible with its older models.

Moreover, there was a strong suspicion that any effort to please the mathematicians would substantially reduce computer performance, all for the sake of seemingly minor details that didn't matter most of the time. A few complained that the existence of such a standard would in fact impede the future development of even better mathematical techniques for accurate computation.

Nevertheless, the late 1970s saw the beginnings of a revolt by a growing number of microcomputer users against the rampant anarchy in the computer world—not only against quirky arithmetic but also many other matters of design and function. There arose an impetus to do something about the tradition of computer manufacturers playing by their own rules and refusing to adopt sensible standards, thus placing unnecessary burdens on programmers and users.

Work on the arithmetic standard was sparked by John F. Palmer, a mathematician at Intel Corporation, a major computer-chip manufacturer in Santa Clara, California. Palmer had persuaded Intel's management to adopt a company-wide arithmetic standard for its lines of microprocessors so that all these chips would be compatible. He argued that it wouldn't look good if computers carrying the same brand of microprocessor—but developed by different groups at Intel's various centers—would sometimes give different answers to identical numerical problems. At the same time, the company could use this as an opportunity to develop the best possible floating-point arithmetic.

Palmer brought Kahan in as a consultant. As a student at Stanford, Palmer had heard several of Kahan's lectures. He had been sufficiently impressed by his arguments to think that Kahan was the right person to specify the requirements for a superior float-ing-point arithmetic.

When competitors heard rumors of these developments at In-tel, they became a little nervous. As Kahan facetiously puts it, they responded with: "Let's slow them down by forming a committee." The effort to develop a binary floating-point standard was under way.

The committee formed under the auspices of the Institute of Electrical and Electronics Engineers to formulate the standard ended up taking an unusual course. Instead of surveying what had been done in the past and coming up with something close to past and present practice as is customary in such efforts, the commit-tee, with about seventy voting members representing a wide range of companies and institutions, chose to start nearly from scratch.

It proved a stormy course to take. Because there was no com-mon, uniform, generally understood doctrine on how this ought to be achieved, committee meetings often rang loudly with heated arguments over one approach versus another to specific issues and concerns. Much of current practice appeared more a matter of habit and taste than anything else, but a number of the experts on the committee clung fiercely to their particular beliefs. Many were skeptical that a consensus and a reasonable standard could be achieved at all. Some worried that the committee was going off to define something that nobody would build.

People were almost literally jumping up and down on the table, saying one couldn't possibly implement the recommended func-tions at a reasonable cost. But Intel cut this debate short when its 8087 microprocessor arithmetic chip debuted in 1980 and demon-strated in a very concrete way that the principles in the proposed standard could be built into an integrated-circuit chip at a reason-able cost and without unduly degrading the chip's performance.

Other companies started to pay attention even before the stan-dard was finally approved. One factor that made its acceptance pos-sible at the hardware level was the steadily increasing sophistication of the chip-making process, which allowed chip makers to pack more and more circuitry onto paper-thin wafers of silicon to accommodate the extra switches and gates needed for the new arithmetic.

Despite several counterproposals, persistent opposition, and considerable skepticism, the committee in the end got away with its sweeping reform. After ten drafts, the standard became official in 1984. By then, several companies were already adhering to it in various products. It was a significant achievement, representing one of the rare occasions when the hardware-producing community decided that, by and large, the interests of the applications programming community should take priority over its own interests.

The IEEE standard for binary floating-point arithmetic is a carefully integrated collection of ideas and techniques for specifying how a computer should do arithmetic. In addition to instructions for multiplication and other operations, it includes mechanisms for warning computer users that in attempting certain operations, they may be treading on thin ice. It even has something called "not a number," which can be applied to the result of a forbidden operation, such as finding the square root of a negative number or dividing by zero. By using such a designation, the computer can flag danger zones, making it easier for the computer user to track down potential trouble spots in calculations.

Of course, the standard doesn't guarantee that all numerical programs and procedures will produce correct results. But the standard makes it much easier to understand the consequences of each mathematical operation and to estimate the accuracy of a given calculation.

"We live in an imperfect world, and this standard is no exception," Kahan and Palmer noted in 1979. "Absolute safety, if attainable, would have cost more than its worth. This standard was designed to yield, for a slightly increased implementation cost, a considerably safer system with greater capability . . . than typical floating-point systems."

**IN 1989,** Kahan was honored for his work in "making the world safe for numerical computation." He received the A. M. Turing Award, the most prestigious technical award in computer science, presented by the Association for Computing Machinery. Named for the British mathematician whose work in the 1930s and 1940s formed the basis for much of the development of computer science, this award honors contributions of lasting and major technical importance to the computer world.

The award citation particularly emphasized Kahan's key role not only in solving a problem but also in seeing that the solution was adopted. His was the unique achievement of having numerical analysis significantly influence the real-world design of computers. As a result of his deep knowledge of software and of his devotion to the vision of providing computer systems that can be safely used for numerical computation, Kahan has had a lasting impact that will affect virtually everyone who uses computers in the future.

In a sense, the award represented a tribute to Kahan's perpetual crankiness. He had fretted about the intimate details of computer arithmetic for more than three decades, and he had managed to move the huge bulk of the computer community toward cleaning up a key problem in numerical computation, one affecting not only the science and engineering communities but also the business and finance sectors.

Although most computer manufacturers and computer-chip makers now build their arithmetic "engines" according to the IEEE standard, there are still some major holdouts. These include IBM 360/370 computers, VAX computers manufactured by Digital Equipment Corporation, and Cray supercomputers. Overall, however, Kahan concedes that the degree and speed of acceptance of the standard at the hardware level has greatly exceeded his expectations and those of the members of the original standard-setting committee.

The situation isn't quite as settled at the software level, particularly in the computer programs known as compilers, which translate the instructions of computer programs written in a "high-level" language such as FORTRAN or Pascal into simpler, more basic instructions that the machine itself can understand. Writing a compiler is an exacting, time-consuming chore, even for the most talented of the programmers who undertake this essential task.

At Apple, where a few of Kahan's former students have found positions, company programmers committed themselves to writing compilers for Macintosh computers that conformed to the IEEE standard in nearly every detail. Ironically, Apple's own management initially failed to appreciate the great advantages such an approach created for Macintoshes over other desktop computers used for scientific and engineering applications. Instead of building high-performance computers that engineers and scientists could use, the company concentrated on machines designed for other markets and applications, which generally didn't require quite as much numerical computation.

At most other companies, the steady gush of new computer designs, or architectures, keeps compiler programmers so busy writing software for the new machines that they have little time and energy to devote to cleaning up the details of the arithmetic used in the compilers, Kahan complains. "Compiler writers have an unenviable task," he says. "They must reconcile sometimes capricious hardware design to the demands of applications programmers, and no sooner do they get a compiler written than the hardware has changed, and they must rewrite it. So it becomes extremely difficult to refine their product and to build upon experience to suit the needs of the application. In short, we're running as fast as we can to stay in the same place, and I'm not sure we're staying in the same place."

Moreover, because of intense competitive pressures, the fate of a new computer model often hinges on how quickly it performs certain benchmark tasks. Compiler writers spend a great deal of time fine-tuning their programs to accomplish these particular tasks as quickly as possible, sometimes to the detriment of overall performance. Again, floating-point arithmetic falls by the wayside, partly because most people who want to do numerical computation need very few of the special features in the IEEE floating-point standard.

On the other hand, Kahan notes, there is no feature that isn't needed by somebody. That makes it hard to establish a market demand for truly clean arithmetic; too few users have a sufficiently broad perspective to understand the full requirements that an effective compiler should meet.

Nonetheless, things have gotten better over the years. Today's calculators generally have fewer obvious quirks than those of a decade ago. New microprocessors built to the IEEE standard have more reliable arithmetic than earlier models, though faulty compilers can sometimes foil a programmer's intent and subvert a computer program into delivering wrong answers. The miraculous web of approximate arithmetic is more important than ever in application after application, and now many of the tiny but maddening uncertainties that typically accompanied such calculations have been alleviated.

Even with a standard, however, there's plenty of room for mistakes—human and otherwise—in the explosively expanding realm of computing, and these errors are exacerbated by the use of teams of different computers yoked together to solve large, complicated scientific and engineering problems. "This means that frequently, when something bad happens, it happens in an environment where so many things are going on that by the time you find out that something bad has happened, you no longer know where it happened," Kahan contends.

On the evening of the celebration of his impending sixtieth birthday, Kahan had a chance to reflect on his career and on the influence he has had on the small group of graduate students who managed to survive his requirements and to leave Berkeley with a doctoral degree. Kahan talked about perfection, and he described graduate study as the one chance in a lifetime when one could take the time to try to get close to perfection. That was the ideal to which he drove his students.

It is the absolute necessity for clear thinking that obsesses Kahan. In the face of a complex world, he says, "what we have to do is find more economical ways of thinking, and economy of thought is something that man has achieved—when he has achieved it—only through millennia of refinement, by gradually coming to understand things better."

He continues, "It's not that I'm asking people to think clearly all the time. Almost no one does that. What I'm asking instead is that we appreciate that there is such a thing as clear thought, and that we should yearn for it. You should look at your own thoughts and wonder perhaps whether you could clarify them a little bit by going over them again."

# CHAPTER 7

# Absolute Proof

---

**A SLEEK,** powerfully contoured sports car screams to a halt. Its roaring engine cuts back to an insistent purr, then smoothly quits. The door opens and closes with a solid, satisfying *thunk*.

There's nothing here that distinguishes this new model from sports cars of previous generations. The real differences are concealed under the hood, where more than a dozen microprocessors control the injection of fuel into the engine's cylinders, adjust the car's brakes and suspension system, and even diagnose ailments, signaling problems to the driver.

The microprocessors that have become increasingly common in automobiles are not greatly different from those that do the arithmetic, make comparisons, and perform other basic operations in desktop computers. Acting as logic engines, such devices interpret the instructions of a computer program and send the proper signals in the correct order to different parts of the chip to accomplish the required results. In effect, a microprocessor behaves like a huge array of miniature switches, with each operation tripping some switches and resetting others. Out of this logic emerges the correct answer to an arithmetic problem or the appropriate response to a bumpy road.

In the form of fingernail-sized slices of minutely patterned silicon, encased in plastic and mounted on spidery legs, microprocessors reside in an astonishingly wide array of habitats from kitchen appliances and telephones to missiles and medical devices. With their advantages in economy, performance, and flexibility, such programmable electronics are steadily increasing their domains, venturing more and more into the territory of critical systems, where design flaws could have fatal consequences.

In the early 1980s, Great Britain's Ministry of Defense, like military establishments elsewhere in the world, had to face the problem of assuring the reliability of a growing array of high-tech weapons triggered or guided by computers. Even at that time, the microprocessors available for military computers were extremely complicated devices, containing tens of thousands of active components—and sometimes more. Inevitably, most microprocessors had both logical and physical defects. Few were good enough for applications in which users needed guarantees that the devices would function properly under all circumstances.

Extensive experience with microprocessors over the years has taught engineers that faults are practically unavoidable, especially in newly introduced commercial products. For example, in late 1989, when customers began testing Intel's i486 microprocessor, they discovered a number of errors in the way the chip performed certain types of arithmetic. To correct the mistakes, the company had to revise one of the stencils, or masks, it uses to print the layers of microscopic circuit features on its silicon chips, delaying product shipments by two weeks. Additional faults were discovered and corrected in subsequent editions of the chip, which since then has become the workhorse of desktop computing. Although Intel's testing procedures had improved considerably over the years, the company still could not catch every error in a chip with 1.2 million components before it was put into production and use.

A similar problem surfaced in 1994 soon after the introduction of Intel's Pentium microprocessor chip. Users of personal computers based on the chip found that for certain types of division problems involving floating-point operations, the microprocessor generated the wrong answers (see Chapter 6).

The chip error caused a considerable furor. In response to the concerns and fears expressed by many users, Andy Grove, Intel's president, issued a statement outlining the company's position:

> The Pentium processor was introduced into the market in May of '93 after the most extensive testing program we at Intel have ever embarked on. Because this chip is three times as complex as the 486, and because it includes a number of improved floating-point algorithms, we geared up to do an array of tests, validation, and verification that far exceeded anything we had ever done. . . .
>
> We ramped the processor faster than any other in our history and encountered no significant problems in the user community. Not that the chip was perfect; no chip ever is. From time to time, we gathered up what problems we found and put into production a new "stepping"—a new set of masks that incorporated whatever we corrected. . . . After almost twenty-five years in the microprocessor business, I have come to the conclusion that no microprocessor is ever perfect; they just come closer to perfection with each stepping. In the life of a typical microprocessor, we go through half a dozen or more such steppings.

Grove went on to point out that the division errors typically occurred so rarely—just once in perhaps nine billion division problems involving decimals of more than five digits—that most users had nothing to worry about. Only those few using Pentium-based computers for heavy-duty scientific or engineering calculations needed to replace the original chips with repaired ones.

This response didn't reassure everyone. Many were concerned about the fact that the company had elected not to notify computer users of the bug when it was first found. Though relatively inconsequential for most applications, the error still mattered as an issue of principle. Of course, previous experience with such problems in both hardware and software has meant that cautious scientists and engineers generally check crucial computer calculations by redoing them using different computers and software.

To build a sufficiently reliable chip to meet its needs, the British Ministry of Defense turned to one of its own research centers, the Royal Signals and Radar Establishment (RSRE) at Malvern in central England. Known for its expertise in radar technology and semiconductor physics, the laboratory also had an extensive research program devoted to high-integrity computing. Since the early 1970s, research center staff had expended considerable effort investigating ways of using mathematical methods for drawing up precise specifications of what a computer program is supposed to do, then ensuring that the program is written in such a way that one can "prove" the resulting software meets the specifications exactly. These mathematically based, "formal" techniques are part of a process known as program verification. In one project, RSRE personnel used formal methods to analyze samples of code drawn from NATO's inventory of military software and found that one in ten fragments contained errors in logic, some serious enough to cause loss of an aircraft or some other machine.

In 1983, three members of the RSRE staff, John Cullyer, John Kershaw, and Clive Pygott, accepted the challenge of applying formal methods to the design of a computer chip. Their hope was to design a new microprocessor and verify its correctness right down to the individual logic gates that open and shut during chip operations—to prove indisputably that they had a truly bug-free chip. Cullyer in particular saw this effort as a crucial step toward creating high-integrity computer systems for both military and commercial applications. He envisioned such a chip used for controlling aircraft, operating medical equipment, securing chemical and nuclear plants, and performing other critical functions.

The result was the VIPER (verifiable integrated processor for enhanced reliability) chip. Compared with other microprocessors then on the market, the VIPER chip was relatively unsophisticated and had limited capabilities. It was slower. It responded to fewer different commands than engineers typically use to program other types of microprocessors. It would simply halt if it got into trouble—for example, whenever a calculation produced a number that exceeded the chip's capacity.

Its key feature was its simplicity, which made it possible to consider undertaking the enormous task of mathematically verifying its logic. "A more complex architecture would have taken the proofs beyond the current state of the art," Cullyer remarked in 1988. Despite the simplifications, however, verifying that the chip worked as advertised proved a monumental, perhaps overambitious, enterprise.

Simply testing all possibilities was out of the question. For example, checking addition by trying all possible pairs of numbers and examining the output to make sure each answer is correct takes too long. There are far too many possibilities. The same difficulty confounds the checking of other microprocessor operations. Even if it were possible to do billions of such tests per second, who could check the answers to make sure they are all correct?

Formal methods furnish a potential shortcut. These involve mathematical techniques that resemble the processes of rearranging equations to show that two superficially dissimilar expressions are in fact the same. Designers prepare a set of specifications that describe the state of the chip's components before and after every operation. To translate these specifications into a precise, complete, correct description of how the chip performs its duties, they can then use techniques similar to those employed in proving mathematical theorems. But it's an onerous, tedious task, involving a great deal of meticulous bookkeeping.

It proved impossible for the RSRE team to achieve a direct proof of correspondence between the specifications and the logic on the VIPER chip itself. Instead, they had to settle for a proof in the form of a chain of reasoning linking four separate levels of abstraction. At the top level, the team members specified the overall changes in the chip's state that would result from each of the instructions in the limited set available to programmers. The next level contained more details about what happens when the chip executes a given instruction. The third level corresponded to a block diagram showing the microprocessor's major components and specifying their intended behavior. The lowest level supplied the details of how individual gates worked, providing the final link between the overall design and the patterns that automated equipment would use to construct masks for fabricating the chip.

The RSRE group recognized that no formal demonstration of correctness was truly possible for the chip. "Physical devices wear out and break down, and no amount of formal logic can guarantee immunity," Kershaw said in 1988. "Ultimately physics makes its own rules."

Because formal methods are not particularly helpful at the level of electrons and transistors, the researchers realized that the best safeguard in the long run would be to repeat the entire design in as different a way as possible. In the interim, the team made special efforts to design the VIPER chip in such a way as to guard against hardware breakdowns.

Two chip manufacturers, using different fabrication processes as insurance against production errors, provided prototype VIPER chips in late 1986. A dozen laboratories began evaluating the device for possible use in battlefield tanks and in other military and commercial settings. By the beginning of 1988, the Ministry of Defense was aggressively marketing the VIPER chip for commercial applications. A major selling point was this device's impeccable pedigree as the first commercially available microprocessor with a mathematically verified design. But the fine print in accompanying reports from the RSRE team indicated that the necessary proofs were at that time still incomplete. Crucial links were still being forged.

Nonetheless, a number of commercial firms obtained licenses

to use and sell the technology, and one customer appeared. The Australian National Railways Commission was interested in using VIPER chips in its signaling system at automatic train crossings.

Then came the bombshell. The National Aeronautics and Space Administration in the United States was one of the organizations that had been chosen to test the prototype VIPER chip. Most of the other organizations evaluating the chip had largely restricted their comments to deficiencies in the chip's performance. NASA had decided to go deeper and examine the claims made about the chip's design. It asked experts at Computational Logic in Austin, Texas, a company developing its own formally proved microprocessor, to evaluate the methods Cullyer and his colleagues had used to verify the VIPER chip's design.

The final report to NASA identified major gaps and a number of "blatant errors" in the purported proofs. Only the link between the top two levels had been established properly. All other links were incomplete or informal. In other words, although the VIPER chip design had been intensively simulated and informally checked, it had not been proven. The authors of the report concluded, "We are not convinced that RSRE researchers have formally verified the gate-level implementation of VIPER."

They added, "We are optimistic that the problems uncovered in the VIPER effort can be overcome, and that this hard-won experience will benefit future work." Cullyer, who by then had left RSRE for a position at the University of Warwick, saw the report and conceded the inadequacies in the verification procedure. The RSRE team acknowledged that "more remains to be done, both to build up confidence in the existing VIPER design and to develop new techniques of design and verification which avoid the limitations of present methods."

Avra Cohn, a formal methods expert at Cambridge University, had expressed similar concerns earlier. She had attempted to apply automated methods to the verification of the VIPER chip logic but had given up after reaching seven million reasoning steps in trying to establish the low-level links. She had also become worried about the way the chip was being marketed and described. In her view, with large gaps still present in the proofs, there was no way to guarantee that the VIPER chip was actually any better than other commercially available microprocessors.

Beyond the completeness of the proofs, there were two additional, serious gaps that no amount of formal logic could ever bridge, Cohn insisted. There is no way, using mathematical techniques, to guarantee that a designer's intentions are accurately or completely expressed in a system's specifications. There is also no formal way to match mathematical gate-level descriptions of a chip with the actual physical device. In other words, a proof would apply only to the blueprints, not to the device itself.

The critical NASA-sponsored report and Cohn's comments effectively killed the VIPER chip's commercial prospects. Alleging that the Ministry of Defense had negligently misrepresented the VIPER chip's status, one small company called Charter Technologies Limited, which had been drawn into the VIPER chip program, sued the ministry. But the firm went bankrupt before the case could come up in court.

If the issue had gone to court, lawyers and judges would have been forced to grapple with the fundamentals of mathematics and logic. What are the limitations of formal methods? What does "proof" really mean in the context of chip design? Indeed, the VIPER design had been subjected to an unprecedented degree of scrutiny, including exhaustive testing, checking, simulation, and mathematical analysis, and no bug had ever been found in the final product.

"The VIPER episode reminds us that 'proof' is both a seductive word and a dangerous one," comments Donald MacKenzie, a sociologist at Edinburgh University who is studying the nature of proof as it arises in university mathematics departments and in industrial situations. "We need a better understanding of what might be called the 'sociology of proof': of what kinds of arguments count, for whom, under what circumstances, as proofs. Without such an understanding, the move of 'proof' from the world of mathematicians and logicians into that of safety-critical computer systems will surely end up, as VIPER nearly did, in the courts."

AT ISSUE IS the notion of proof in mathematics. Defined loosely, a proof consists of a sound, logical argument from a set of fundamental principles, or axioms, that establishes the truth of a statement or theorem. But how does one know when an argument is sound and logical?

In 1979, in a paper with the title "Social Processes and Proofs of Theorems and Programs," Richard DeMillo, then a computer science professor at the Georgia Institute of Technology, and his colleagues Richard Lipton and Alan Perlis, strongly argued that the lengthy chains of dense, logical formulas characteristic of mechanically generated program verifications were not actually *mathematical* proofs. They contended that proofs of theorems in mathematics depend for their acceptance on a *social* process, which is absent from mechanical verification of programs.

They wrote: "A proof is not a beautiful abstract object with an independent existence. No mathematician grasps a proof, sits back and sighs happily at the knowledge that he can now be certain of the truth of his theorem. He runs out into the hall and looks for someone to listen to it. He commandeers the blackboard. . . . Mathematical proofs increase our confidence in the truth of mathematical statements only after they have been subjected to the social mechanisms of the mathematical community."

Taking an even more controversial stance, DeMillo and his colleagues then argued that the necessarily lengthy, formal verification of software corresponds not to the working mathematical proof, which rarely ever goes down to the level of fundamental axioms, but to the largely imaginary logical structures that mathematicians assume underlie their work and give it validity. That puts formal verification into a special category of proof, one that is too unwieldy for practical use. The proofs would be too long and boring. The statements implementing such proofs would generally end up at least ten times as long as the code itself, giving plenty of scope for introducing human error. And few people would actually take the trouble to go through the tedious process of checking the proof after it was completed. Hence, the crucial social mechanism for winnowing out the truth is missing from program verification.

However, this contentious argument failed to distinguish between the social process of checking proofs and the actual truth of the theorems themselves. Whereas social processes may be crucial for determining which proofs the mathematical community takes to be valid or which theorems it takes to be true, they do not determine their absolute validity or truth.

In later defending the VIPER effort, Martyn Thomas, who heads the British software company Praxis, took a slightly different tack: "We must beware of having the term 'proof' restricted to one extremely formal approach to verification. If proof can only mean axiomatic verification with theorem provers, most of mathematics is unproven and unprovable."

He added, "The 'social' processes of proof are good enough for engineers in other disciplines, good enough for mathematicians, and good enough for me. If we reserve the word 'proof' for the activities of the followers of Hilbert [a nineteenth-century mathematician bent on putting mathematical reasoning on more solid ground], we waste a useful word, and we are in danger of overselling the results of their activities."

Even as the VIPER chip was edging toward its demise, doubts about the efficacy of formal methods were stirring up the computer community in the United States. In a lengthy article titled "Program Verification: The Very Idea," published in the September 1988 issue of the *Communications of the ACM*, James H. Fetzer, a philosophy professor at the University of Minnesota in Duluth, attacked the very notion that program verification can succeed "as a generally applicable and completely reliable method for guaranteeing the performance of a program."

Fetzer's argument was that the proclaimed aims of program verification are, even in principle, impossible to achieve. The source of the difficulty lies in the fact that programs run on real computer systems, involving physical components that can't be precisely characterized and that can behave unpredictably. A mathematical demonstration of a program's correctness does not guarantee that it will still function correctly when run in real time on a physical machine. It's the same difficulty that stymies the formal verification of a chip design.

"There is little to be gained and much to be lost through fruitless efforts to guarantee the reliability of programs when no such guarantees are to be had," Fetzer concluded. In essence, he was agreeing with the general conclusion that DeMillo and his colleagues had arrived at in 1979, for what Fetzer described as, "the wrong specific reasons."

"The limitations involved here are not merely practical: *they are rooted in the very character of causal systems themselves*," Fetzer emphasized. "From a methodological point of view, it might be said that programs are conjectures, while executions are attempted— and all too frequently successful—refutations."

Fetzer's article provoked a storm of protest. Many in the program-verification community, which depended on the U.S. Department of Defense and the supersecret National Security Agency for the bulk of its funding, looked on the paper as a direct, personal attack, and they replied in kind.

The most vehement response came in a letter from ten prominent computer scientists, led by Susan L. Gerhart of the Microelectronics and Computer Corporation. The list also included Peter Neumann of SRI, Victor Basili of the University of Maryland, and Nancy Leveson, then at the University of California at Irvine. In essence, they insisted Fetzer had asserted erroneously that the purpose of program verification is to provide an absolute guarantee of correctness when a program is actually executed on a real computer. No one seriously advocates such a view, they argued.

"The article is ill informed and irresponsible because it attacks a parody of both the intent and practice of formal verification," the group submitted. "It is dangerous because its pretentious and ponderous style may lead the uninformed to take it seriously." In a blizzard of intemperate terms, they accused Fetzer of "distortion," "gross misunderstanding," "misrepresentation," and "misinformation."

And this was the toned-down version of the letter! The original drafts were considerably more harsh and incendiary.

Fetzer fired right back, zeroing in on the difference between absolute and relative verification. "This difference, of course, is critical," he replied, "since the methods of absolute verification are conclusive but cannot be applied to causal systems [like real computers], while those of relative verification can be applied to causal systems but are not conclusive."

Fetzer went on to declare that his opponents had suffered "a complete failure to understand the issues," citing their "intellectual deficiencies" and "inexcusable intolerance and insufferable self-righteousness" and noting that they had behaved like "religious zealots and ideological fanatics." He went so far as to invite them to "volunteer to accompany [programmable cruise] missiles on future flights in order to demonstrate that this is a type of programming that actually does lend itself to the construction of program verifications."

The salvos continued through the letters and comments sections in several more issues of the *Communications of the ACM*. Defenders of program verification readily conceded that absolute verification is impossible, but they insisted that formal methods serve a number of very useful purposes, for example, in locating logical faults at the design stage—before they can cause harm. Formal methods serve as tools for making programs more reliable, partly by getting programmers to pay more attention to details. They are not meant to guarantee absolutely that a program will work perfectly when run on a computer.

A few respondents took Fetzer's side, arguing, for example, that many useful programs are not merely unverifiable but also incorrect, partly because computers can at best only approximate certain types of numbers that come up in calculations (see Chapter 6). Other commentators remarked that the proponents of formal methods aren't always completely honest about the limitations of their methods. Too often, the warning labels are missing, and users apply the methods carelessly. Formal methods have their place, but they shouldn't be overemphasized or oversold, it was argued.

Part of the passion evoked by Fetzer's arguments came out of a genuine, and perhaps justifiable, fear that his pessimistic conclusion about the value of "absolute" program verification would be misunderstood by outsiders as an attack on formal methods in general. They especially did not want funding organizations such as the Department of Defense believing that their money was going to waste.

The argument sputtered on. In the end, the debate didn't settle much of anything, and Fetzer's arguments did not derail program verification. Formal methods in their myriad forms continued to be strongly advocated by a significant proportion of the computer community, though not widely used in practical situations. Their proponents are now perhaps less extravagant in their claims and less casual in their use of the emotionally charged word "proof." As Cohn had pointed out in the VIPER case, the very language of proof and verification can convey a false sense of security, particularly among those who are unaware of precisely what computer scientists and software engineers are up to.

Mathematical concepts underlie the efforts of Harlan Mills (see Chapter 4), David Parnas (see Chapter 3), and other computer experts, who apply formal methods for writing defect-free software and for checking programs. But these experts also understand the limitations of formal strategies. At the same time, several groups have gone ahead with efforts to "prove" the correctness of various types of microprocessors, with varying degrees of success. Others have plunged ahead with attempting formal verification of programs.

Much of the effort to use formal methods may be wasted, however, if the entire problem isn't thought out carefully. "Often, I find groups building or using mathematical models based on unrealistic and unproved assumptions," Leveson says. "I also see projects where the majority of the effort is expended in attempting to prove theoretically that the system has met a stipulated level of risk when it would have been much more profitably applied to eliminating, minimizing, and controlling hazards."

Peter Neumann has long argued that checking specifications to make sure they are consistent with what is expected of a system may be more important than verifying programs to eliminate bugs. His own experience in developing secure systems has shown that systematic procedures for checking specifications can uncover design flaws. The real benefit of verification schemes is not in proving anything absolutely but in pinpointing defects along the way.

Formal methods also figure prominently in the deliberations of a number of regulatory and standards organizations that are investigating how to certify safety-critical systems, ranging from the U.S. Federal Aviation Administration to Canada's Atomic Energy Control Board. A variety of draft standards and regulations now in circulation mandate the development and use of mathematically based techniques for assuring safety and security.

A great deal more remains to be done. A recent independent study of the use of formal methods in the development of critical systems concluded that "formal methods are applicable to industrial software development and should not be ignored by industry." But the study also concluded, "the industrial application of the technology is constrained by the sophisticated nature of the technology, negative connotations, and the small cadre of experts."

"There's simply an awful lot to know to build a complex system," says Susan L. Gerhart, now director of the Research Institute for Computing and Information Systems at the University of Houston at Clear Lake and one of the authors of the 1993 study. "Specialists may understand how to control complexity of some aspects of a system, but total knowledge is beyond any individual, beyond most projects, and poorly organized."

Part of the answer lies in developing an "applied mathematics for software engineering," she contends. "What is needed is a massive infusion of analysis capability into routine software engineering." In other words, software engineers need the ability to evaluate and predict the quality of a system as it is being developed. Although great strides have been made in formal methods, these techniques remain too limited and inflexible for thorough analyses of computer systems.

One reason for the slow rate of progress, she maintains, is that there hasn't been any incentive for people to change conventional practice. "The discipline barriers are very strong—performance people and correctness people just don't talk much," she notes. "The U.S. cultural mistrust of mathematics and infatuation with [artificial intelligence] and heuristics works against a broad-based change."

C. A. R. Hoare, a pioneer of program verification at Oxford University, pointed out in a 1986 article that, in principle, a programmer could learn to write completely formal proofs, and a computer could be programmed to check those proofs. However, "the labour of constructing proofs with sufficient formality for a machine to check them has turned out to be excessive," he admitted. For the time being, "the most effective way of checking proofs is to submit them to the gaze of another programmer or mathematician. . . . The checker then joins the programmer in taking responsibility for the correctness of the program." At the same time, tests are necessary to check the adequacy of the original specifications and the general assumptions on which they were based.

"The principle of diversity is fundamental to improving the reliability of programs," Hoare maintained. "The people who check proofs should be independent of those who construct them. Those who design test regimes should be independent of those who design the objects undergoing the tests. Finally, for ultimate confidence, it would be ideal to have two independent proofs, preferably using different methods of proof. It is by independent experiment that the laws of physics are confirmed; and even mathematicians are happier with fundamental theorems that have been proved more than once."

The two main choices that programmers have for ensuring software reliability suffer serious limitations. Applying the formal methods of program verification, they can prove that a given program works correctly for all possible inputs. But that's usually a very difficult, if not infeasible, proposition. Moreover, even if the proof is correct, it makes a statement only about the program as it is on paper and says nothing about the correctness of the compiler software that must translate the program statements into instructions a computer can handle, and nothing of the computer hardware itself.

Alternatively, programmers can test software by feeding it various inputs. That, however, leaves no assurance that the tests cover all types of inputs a program is likely to encounter. Thus, a user has no guarantee that the program output is correct for a particular input. To complicate matters, during testing, the test answers generally must be compared with the "correct" answers, which often come from other computer programs.

IN RECENT YEARS, computer scientists have started to explore a new approach that may offer an attractive supplement to program testing and program verification. The approach emerges from theoretical computer science, a field often regarded by programmers and other computer professionals as far removed from, or even irrelevant to, practical concerns. Its key insight is that verifying the correctness of an entire program isn't always strictly necessary. It may be sufficient for many purposes to check merely that a particular answer is correct.

In other words, program (or result) checking provides a way of ascertaining whether a program, given a certain input, produces the correct answer for that particular input. Conceptually, the process isn't very different from the high-school math exercise of substituting a calculated answer back into the original equation to check that the answer is correct.

The man behind the concept of program checking is Manuel Blum, a veteran computer scientist at the University of California at Berkeley. The notion of program checking originated during Blum's brief 1988 sojourn at IBM's Thomas J. Watson Research Center in Yorktown Heights, New York. The center's main building, a large, low, circular structure crowning a grassy hill, has the feel of a small-town college, minus the bustle and clamor of a youthful student body. The building's dark glass panels, glinting in the sunlight, hint at a refuge for quiet contemplation and the nurturing of innovative ideas.

Blum was visiting his colleague and friend Shmuel Winograd, a computer scientist who works at the center. At one point during the visit, Blum asked, "In your view, what is the most important research problem facing IBM?" Winograd's immediate answer was, "Program testing!"

Actually, Winograd had more than one answer to Blum's question, but it was this particular answer that lodged in Blum's mind, sending it into action. His chain of reasoning started with the observation that students, engineers, and even programmers are all taught to check their work, though they aren't necessarily given good, practical advice on how to do so properly. An engineer, for example, after completing a calculation to determine the load a bridge is to carry, is expected to check whether the calculated answer makes sense. If the resulting number is outside the normal range for the given situation, the engineer can look for an error in the calculation or a defect in the reasoning that underlies the calculation.

In the computer world, numerical analysts and compiler writers have for years inserted messages into their programs warning of potentially troublesome situations, such as dividing by zero, that can lead to erroneous results. What Blum had in mind went much further. His view was that errors could arise in *any* computation, that every computation must be checked in some way, and that warnings should be posted of any problems.

"By allowing the possibility of an incorrect answer (just as one would if computations were done by hand), the program designer confronts the possibility of a bug and considers what to do if the answer is wrong," Blum says. "This gives an alternative to proving a program correct that may be achievable and sufficient for many situations."

Hence, a program should be written incorporating simple segments of code—result checkers—that run in conjunction with the main program and allow it to check itself. A large program may, in fact, have several embedded checkers. The checking then happens every time the program is run. This contrasts with program verification, which occurs before a program goes into use.

"The lack of correctness checks in programs is an oversight," Blum maintains. "Programs have bugs that could perfectly well be caught by such checks." The question comes down to what constitutes an adequate procedure for performing an independent check of a calculation in a computer program. Clearly, simply repeating a calculation is not enough.

Recent advances in theoretical computer science pointed Blum toward a remarkably powerful means of putting his idea into effect. At the root of these advances lies the startling notion of a probabilistic, interactive proof. Unlike traditional methods, familiar to any student who has tried to prove one of Euclid's geometrical theorems by constructing a chain of statements inexorably leading to the necessary conclusion, the new concept of probabilistic, interactive proof relies on randomness and the interplay between a "prover" and a "checker" to achieve a practically unassailable proof.

Blum turned to the work of computer scientists Sylvio Micali, Oded Goldreich, and A. Wigderson, who had used the concept of probabilistic, interactive proofs as the basis for the wonderfully counterintuitive notion of a "zero-knowledge proof." Such a scheme permits someone to persuade others that a particular theorem he or she has proved is true without giving away anything about how to go about proving the theorem itself.

"What is new and very different is that one can convince somebody of a proof's validity by using coin flipping and interaction," Blum says. "I'm excited about that because I think theoretical computer science has introduced a new paradigm here—a paradigm proving to be very powerful for many different reasons and purposes."

Taken into the world of commerce and sensitive data, such schemes can be adapted to create secure, cryptographic methods for identifying computer users, certifying computer transactions, and vouching for the integrity of information. Variations on the technique can ensure that contracts are signed simultaneously in different parts of the world. It can be used as the basis for an electronic signature.

Applying the idea to software, "it's possible to get a program to convince you that the answer it gave you is correct," Blum adds. "What you're interested in is either knowing that there's an error somewhere, in which case you would debug the program, or getting strong, convincing evidence that the particular answer you got from the program is correct, even though the program itself may contain bugs."

Blum illustrates the role of interactive probabilistic proofs in software bug detection using the example of determining whether two apparently dissimilar graphs—mathematical constructions made up of sets of points, or nodes, connected to one another by lines to form networks—are really the same. Such graphs may represent, for example, the electrical properties of a circuit and the related pattern of connections fabricated in a silicon chip. Often, these graphs look quite different, and engineers are interested in knowing whether an electrical design (represented by one graph) is correctly implemented on the chip (represented by the other graph).

Researchers have developed a number of reasonably efficient schemes for determining mathematically whether two dissimilar graphs are really the same, and computer programs incorporating those schemes do the actual work. Thus, using a standard, off-the-shelf computer program, an engineer can readily ascertain whether two graphs are really the same. The program checks the two graphs, and it produces a yes or no answer. But is the yes or no answer sure to be correct?

Now the checking part of the program comes into play. If the two graphs really are the same, the checker has a standard, well-known method for confirming a yes answer. If the answer is no, the checker feeds the program one of the two graphs and either a new version of this graph—in which the nodes are randomly relabeled but the connections remain unchanged—or a randomly relabeled version of the second graph.

If the program (and everything else, including the computer hardware) is functioning correctly, then it should give a "yes, it matches" answer whenever it sees one graph paired with its relabeled version, and a "no, it doesn't match" answer whenever it sees this graph and the relabeled version of the other graph. Any other responses would indicate that an error lurks somewhere in the software or hardware.

In case of an error, the program has a negligible chance of "guessing" correctly each and every time, because a checker presents one or the other of these alternative pairings randomly and repeats the process a few dozen times. "There is a chance that the program can fool the checker, but the chances of doing so are extremely small," Blum says.

Although the checker can signal the presence of an error, it does not identify whether the defect resides in the program, the checker, or the hardware, or what the defect is. It's up to the user to decide how to respond to the situation. On the other hand, even a bug-infested program can still generate correct answers, and one readily obtains assurance that those answers are acceptable. One can still get something worthwhile out of a flawed program. It needn't all be thrown away.

"The important thing here is that the checker is basically saying 'yes,' the answer you get can be relied on, or 'no,' the program [or computer system] has a bug somewhere in it," Blum says. In the latter case, "maybe the answer is correct. Maybe it isn't. But you know you can't rely on it."

Comparing graphs is not the only case in which program checking works. Depending on the computation, program checkers can rely on a number of different mathematical schemes for their efficacy. Blum and his collaborators have been studying several alternative methods of formulating program checkers.

In one sense, Blum and his students are going back to the early days of computers when the machines were so unreliable that every calculation had to be checked extremely carefully. Programmers had to build in all kinds of little tests to ensure that the answers were legitimate. William Kahan, Blum's colleague at Berkeley, notes that the trick is having the judgment and experience to know when and where to check for errors. There is not much point in checking for an error that may come up only once in the lifetime of the universe, he says, but there is great value in finding those errors that occur frequently enough to threaten a computer's usability.

Now, whenever Blum writes software or even uses software written by other programmers, he routinely includes checkers in the programs. For example, he wrote a checker for a commercially available calculator program designed to perform certain kinds of arithmetic useful in cryptography. On one occasion, long after he had forgotten his software checker was there quietly doing its duty in the background, Blum suddenly heard a tiny beep signaling that the checker had found a problem.

"It was the most amazing thing in the world to hear it suddenly go off to say it had discovered a bug," Blum says. "I had forgotten all about the checker."

Although the commercial calculator program handled many pairs of ten-digit numbers successfully, there was one particular pair on which it had failed. It turned out that the programmer who had written the original calculator program had inadvertently introduced a bug that became evident only under very special circumstances. Blum's checker caught the unexpected mistake.

"The fellow who had written this program had a very clever way of speeding it up, and in the process of speeding it up, he introduced a bug," Blum remarks.

Programming, however, is not Blum's primary interest. He's a theoretician, and he and his collaborators are rigorously developing the theory underlying the way checkers work, extending their domain to larger programs and additional problems. Blum shares his work by speaking at universities, various other institutions, and occasional meetings, quietly spreading the word within the theoretical computer science community. The interactions with other researchers engendered by his talks, along with interactions with graduate students working with him, play a large role in his efforts to refine, extend, and elucidate the basic concept of program checking.

In 1993, however, the extratheoretical world caught up with him. Unknown to Blum, the Department of Defense in November of that year issued a request for innovative research proposals, which specifically mentioned Blum's work on program checking.

Blum started receiving telephone calls from companies interested in obtaining additional information.

Within a month, Blum had become a consultant to the Hughes Aircraft Company, which is interested in putting checking programs into its avionics software. The company has collected more than forty examples of software errors that extensive testing procedures had failed to catch before the programs went into use in test aircraft. "You would like to get the bugs earlier," Blum says. He will be part of an effort to build checkers to prevent such embarrassing failures in the future.

The prospect of looking at computer programming in an industrial setting delights and excites Blum. It's a chance to explore new, different kinds of thinking, and that fits well with his lifelong desire to understand the nature of human thought—and its inevitable companion, human error.

The human brain works its wonders silently. There's no clashing of gears, no clicking of keys, no electrical hum. Just the exquisite quiet of human thought, hidden away from any onlooker. Out of this invisible realm comes the panoply of human memory, imagination, logic, and creativity.

Our immense processing power is packaged in a lump of wet tissue weighing about three pounds and occupying a space only a little larger than a quart container. Its basic unit is a spindly cell called a neuron, and it contains one hundred billion of these components—roughly equal to the number of stars in our galaxy. A typical neuron may make as many as ten thousand connections with other nerve cells to create an exceedingly complicated network that somehow coordinates a wide range of activities and functions, yet allows a remarkable degree of local autonomy and control. The brain can also store an enormous amount of information. Estimates of its capacity range as high as ten trillion megabytes, a figure that exceeds the combined memory needs of all desktop computers now in use.

With such processing and memory capabilities, it should come as no surprise that the brain's densely packed constituents consume a lot of energy. For its weight, brain matter is the most biologically active tissue we have. Its activity shows up as fluctuating electrical and magnetic signals, which reflect transport of chemicals into and out of cells and from cell to cell. These signals provide a window through which biomedical researchers can probe the system's overall performance.

Even under the best of circumstances, our brains don't function perfectly. We do forget. We can be fooled. We make mistakes. Although complete failures rarely occur, neural systems often suffer local faults. But the plentiful neurons and their myriad interconnections apparently provide sufficient redundancy, flexibility, and stability to keep local failures from affecting the system's overall performance.

It is this system that dreams of what may be, that strives to achieve those visions, that creates the cultural and technological infrastructure to sate those desires. It produces software, and to do so, it forces itself to think logically—an unnatural act in a world where information is invariably incomplete or limited and data are faulty or ambiguous. In such an environment, human decisions must be based not on cold, clear logic but on intuition, common sense, and experience.

Software is the mental battleground where human needs, predilections, and desires contend with brittle logic, mathematical rigor, and inhuman precision. Inevitably, there are bugs.

# CHAPTER 8

# Human Error

LAUNCHED ABOARD a massive Proton rocket on July 7, 1988, from the Baikonur Cosmodrome on the bleak, brown plains of Soviet Central Asia, the *Phobos I* spacecraft had a rendezvous with Mars. Named for the larger of the planet's two diminutive, misshapen moons, the space probe was packed with an array of sophisticated instruments for studying Mars. Like the *Voyager* spacecraft that flew by Jupiter and the other giant planets of the solar system, *Phobos I* carried computers to control the probe and operate the instruments.

By August 29, *Phobos I* had traveled more than twelve million miles of its 111-million-mile trek to Mars. That night, a ground control operator at the mission control center near Moscow prepared to transmit a long message via radio to the spacecraft. This message consisted of a list of different digital commands and instructions, which the probe's onboard computer was to implement during a brief period when communication with Earth was interrupted. Normally, the center used a special computer to check all commands sent to the spacecraft, but on this night, the watchdog computer wasn't working. Moreover, in violation of control center rules, a backup human operator was not on duty to check the controller's work.

The controller decided to send the commands anyway. Unfortunately, he had inadvertently omitted a single letter in one of the instructions. The spacecraft's computer interpreted the erroneous signal as a command to stop locking the spacecraft's guidance system on its celestial landmarks, the sun and the star Canopus, and shut off the thrusters of the orientation system. Without these reference points and the thrusters to keep itself properly oriented, the spacecraft began to drift from its correct orientation. The pressure of the sun's light on its large, saillike solar panels gradually overturned the craft, and these vital, energy-absorbing solar panels ended up facing away from the sun for sufficiently long periods to interrupt the recharging of the batteries aboard the vehicle. Within forty-eight hours, the batteries had died and the probe's systems had frozen. By the time the fault was discovered, not even the spacecraft's receiver had sufficient power to respond to new commands from the control center. As one Soviet space official said at the time, *Phobos I* had fallen into a "deep, lethargic sleep" from which the spacecraft never awakened.

Was the loss of the spacecraft mostly a matter of bad luck? Was it mainly human error? Or was bad design also to blame? Some news reports focused on the human controller's incompetence. Although the controller was the one who eventually tracked down the error, he was barred from participating in the subsequent operation of *Phobos II*, the companion Mars-bound spacecraft launched two weeks after *Phobos I*.

One can argue that design decisions made during the development of the spacecraft contributed to the problem. The craft's onboard computer should have immediately rejected the erroneous, self-destructive instructions, but it didn't. Perhaps the use of a computer language or command structure that could easily be misconstrued contributed to the debacle. Clearly, the humans in the loop failed to follow the procedures laid down for them. And notable successes in previous years had made the Soviet bureaucrats and engineers running the space program complacent.

It's possible that computer problems also played a role in silencing *Phobos II* before it could complete its mission. In January

1989, engineers discovered that out of three identical processors on board the spacecraft, one had already died and a second had a penchant for malfunctioning occasionally. This was potentially a serious problem because the system's logic was based on a "vote" by each processor, with the majority opinion adopted as the final decision. Although this made sense if all three processors were functioning properly, there was no provision for a situation in which two units failed. The remaining processor, even if it were in perfect condition, could do nothing. Its vote could never outweigh the absent votes of its dead companions, and the logic could not be changed.

*Phobos II* made it to Mars, but while orbiting the planet, it lost contact with Earth on March 27. Officials of the Soviet aerospace industry, which had developed and built the spacecraft, never conducted a full investigation of the failure, admitting only that one final, faint message had come from the spacecraft. This last gasp suggested that this vehicle, too, had lost the capability to orient itself, and its batteries had run down. Why it happened remained a mystery.

"The aerospace industry, closed to such truth-seeking since its inception, had a well-developed psychological resistance to scrutiny," space scientist Roald Z. Sagdeev recounts in his memoirs. "Its bosses had not only become accustomed to unnecessary secrecy in which they tried to hide actual happenings, they had also learned how to use tricks to avoid responsibility and punishment in the face of failure. The loss of *Phobos II* proved no exception. . . . The Soviets tried hard to hide the facts related to failures, but a system unable to learn from its own mistakes must be doomed."

Stories of "human error" abound in every industry, from nuclear power and aviation to business and finance. As Peter Neumann's Risks forum attests, the computer industry itself serves up case after case.

"People err. That is a fact of life," says Donald A. Norman, a noted expert on the psychology of design and a fellow at Apple

Computer in California. "People are not precision machinery designed for accuracy. In fact, we humans are a different kind of device entirely. Creativity, adaptability, and flexibility are our strengths. Continual alertness and precision in action or memory are our weaknesses."

Furthermore, "the same properties that lead to such robustness and creativity also produce errors," he continues.

Failures involving computers are often attributed to human error. Users of personal computers know how easy it is to mistype a command, to misunderstand the instructions in an opaquely written manual, or to do something that seems to make perfectly good sense but doesn't meet the computer system's exacting criteria. The same kinds of errors can occur with any computer system, whether processing words or controlling an aircraft, and the fault sometimes lies more with the system designers than with the users.

Norman argues that instead of blaming the human who happens to be involved, it would be more productive to identify the system characteristics that led to the incident and then to modify the design, either by eliminating the situation or at least minimizing its impact in the future. He calls for removal of the term "human error" from the vocabulary of computer professionals and for a reevaluation of the tendency to blame individuals.

Would that make "design error" a more common term? Design itself is a characteristically human activity, subject to the same kinds of triumphs and failures as any other human endeavor. Software design is the creation of artifacts that are truly products of the human mind—products that don't exist in any physical sense and that reflect both the astonishing ingenuity and the intricacies, idiosyncrasies, and inconsistencies of the human mind.

**ALTHOUGH IT'S RARELY** explicitly mentioned in conference programs, the notion of human error and what it means in the context of computers figures prominently in a variety of professional meetings devoted to software testing and evaluation, maintenance, reliability, and other vital concerns. Many of these gatherings focus on crucial issues and new research in the development and validation of software designed for aircraft avionics systems, nuclear power plants, medical devices, and other critical applications.

However, compared with attendance at the huge, extravagantly staged computer shows devoted to the unveiling of new commercial products, the annual gatherings of those immersed in computer graphics, or even the conferences on object-oriented programming (the latest fad in software engineering), meetings concerning critical systems and the reliability and safety of software invariably draw much smaller crowds. They attract a minuscule fraction of the vast worldwide computer community. The emphasis on caution in these conferences makes for an uncomfortable fit in an industry that thrives on optimism and revels in rapid, headlong advances in technology.

In 1991, at a conference on software for critical systems, one key paper focused on the kind of ultrareliability required for airliner avionics systems and other critical applications. The European-built A320 airliner and many military aircraft already use computer-controlled "fly-by-wire" systems (see Chapter 1). Because the slightest fault can lead to disastrous consequences, the software for such a computer system must be extremely reliable. The trouble is that testing complex software to estimate the probability of failure cannot establish that a given computer program actually meets such high levels of reliability.

The paper's authors, Ricky W. Butler and George B. Finelli, computer scientists at the NASA Langley Research Center in Hampton, Virginia, took a close look at the testing methods typically used to demonstrate that a computer program meets a certain level of reliability. The traditional method of determining the reliability of a lightbulb or a piece of electronic equipment involves observing the frequency of failures among a sample of test specimens operated under realistic conditions for a predetermined period. Using these data, engineers can estimate failure probabilities of not only individual components but also entire systems. It's the kind of statistical torture testing that companies, consumer groups, and government agencies often conduct to determine the reliability of automobiles, appliances, and other products.

Unlike hardware, however, software doesn't wear out or break. Indeed, nearly all software problems can be traced to faulty human reasoning. Unless they are caught, the errors persist throughout a system's lifetime. That makes conventional methods of risk assessment difficult to apply. The problem is further compounded by the high degree of reliability required for life-critical applications. Historically, manufacturers of aircraft avionics and other systems in which faults could threaten human lives have accepted a reliability level that corresponds to a failure rate of about one in a billion for every hour of operation.

Butler and Finelli demonstrated that techniques often used by software engineers to quantify software risk are too time consuming to be practical when used to assess systems that require extremely high reliability. For example, software design often involves a repetitive cycle of testing and repair, in which the program is tested until it fails. Testing resumes after the cause of failure is determined and the fault repaired. But it generally takes longer and longer to find and remove each successive fault. To establish that a complicated computer program presents minimal risk would require years, if not decades, of testing on the fastest computers available.

The analysis by Butler and Finelli also affirmed that the use of redundancy in computer systems doesn't necessarily guarantee sufficiently high reliability for those systems' use in truly critical applications. For instance, one way computer-system designers sometimes attempt to reduce the risk of failure is to use multiple versions of a program, written by different teams, to perform certain functions. Although each version may contain flaws, it's highly unlikely that all or even a majority of the programs would contain the same error. However, experiments conducted by Nancy Leveson and John Knight have shown that computer programs independently written to do the same thing often contain surprisingly similar mistakes. One reaches the inescapable conclusion that the difficult parts of any design are likely to lead to errors no matter who codes them, and it is precisely those parts that may fail on the same inputs.

In general, there's no feasible way at present to test software and become certain that it has met the goal of ultrareliability. Similarly, although mathematically based "formal" methods can increase one's confidence in software, they cannot "prove" its ultrareliability (see Chapter 7). In some cases, a computer system could actually meet its quality goal, but at the highest required levels of reliability, it would be impossible to demonstrate that it does.

"This leaves us in a terrible bind," Butler and Finelli wrote in their paper. "We want to use digital processors in life-critical applications, but we have no feasible way of establishing that they meet their [ultrareliability] requirements."

They concluded, "Without a major change in the design and verification methods used for life-critical systems, major disasters are almost certain to occur with increasing frequency."

Such pessimism doesn't pervade the computer community at large. More pragmatic software engineers insist that it's foolish to require so much of software that it isn't or can't be used in certain situations unless it's absolutely clear there are acceptable software-free alternatives. Often, eschewing a software component can make a system less reliable than it would be otherwise. Moreover, some would argue there's really not much point in requiring from software a level of assurance considerably more stringent than from other components of a given system.

At the same time, there's an underlying anxiety and uncertainty felt by those attempting to establish software engineering as a true branch of engineering. Their aim is to apply to the process of designing and writing software the same traditions, standards of engineering practice, and discipline that govern mechanical, electrical, or civil engineering. In some sense, programming itself can be thought of as mathematical engineering, for which mathematics provides the underpinnings, just as chemistry does for chemical engineering and physics does for mechanical, electrical, and civil engineering. Beset by horrendous complexity, an all-too-brief history, and widely divergent opinions on the best or most appropriate methodologies to apply to a given problem, software engineers generally view with envy their role models in more established branches of engineering. They point to the apparent ease and success with which bridge builders, power-system designers, automotive engineers, and other engineering professionals ply their trade.

BUT ERRORS—human and otherwise—are not restricted to the software domain. Henry Petroski, head of the department of civil and environmental engineering at Duke University, has long studied the role of failure and human error in engineering design, a recurrent theme in his writings, lectures, and research.

Petroski carefully distinguishes between science and engineering, noting that engineering typically deals with the behavior of objects far more complex than that of the elementary components from which they are made. For example, physics precisely defines the interactions between the atoms that make up a steel beam, but engineers need to know whether the beam will crack or bend under a certain load. Knowing the laws that govern how individual atoms behave doesn't help. Engineers usually can't work at the atomic level. They need empirical rules and guidelines to estimate a beam's strength and rigidity and to judge whether a network of such beams will stand erect as the framework of a building or bridge.

Precisely because engineering is not science, engineers must rely on their judgment—an amalgam of common sense, intuition, and experience—to proceed with their designs and structures. Thus, Petroski says, engineering design is a succession of hypotheses that a certain arrangement of parts will perform a desired function, that an electronic circuit will produce a certain electrical output, or that a bridge will not collapse. Failures show where the engineers' hypotheses are faulty. Although engineers do not intend to learn by making mistakes, it is only from failure that they discover how to advance the state of the art, Petroski argues. Success by itself doesn't provide enough information.

"To design is to obviate failure, and to proceed through the design process is to anticipate all possible ways in which the object or system being designed can fail," Petroski says. "Thus the concept of failure is key to any discussion about design."

"Failure also plays a central role in the major paradox of engineering design: that successful designs tend to evolve into unsuccessful ones that fail, while failures lead to renewed successes," he continues. "This paradox is resolved when one views the design process from the point of view of failure avoidance, and it provides a means of articulating a coherent view of the history of engineering, including its apparent cycles of success and failure."

One of the most famous and spectacular failures in structural engineering occurred in 1940 when the Tacoma Narrows suspension bridge in Washington state collapsed while being buffeted by a stiff wind. Caught in the wind's grip, the pavement rippled and twisted until finally the oscillations proved too great for the structure, and huge sections collapsed into the water below. Captured on film, the graceful bridge's unexpected, untimely demise left an indelible impression, prompting changes in the ways suspension bridges are built.

This wasn't the first time that engineers had encountered major problems in the design and construction of suspension bridges. Indeed, suspension bridges, in which the roadbed hangs from cables, have a long history of large-scale oscillations and catastrophic failure under high and moderate winds. The earliest recorded problem involved a 260-foot-long footbridge constructed in 1817 across the River Tweed in Scotland. A gale destroyed the bridge six months after its completion.

In 1854, high winds completely destroyed the roadway of a 1,010-foot suspension bridge across the Ohio River at Wheeling, West Virginia. An eyewitness wrote that the structure lunged like a ship in a storm, finally crashing into the waters below. There are a number of other examples. In some cases, the bridges didn't actually shake themselves to pieces, but the wind-driven oscillations sometimes grew so large that a traveler crossing the bridge would get seasick.

After 1854, engineers gradually learned how to increase the stability of suspension bridges by making them heavier and stiffer. As their confidence grew with each successive, successful design, they strove for longer, lighter, and more slender bridges. Driven by the need to keep costs as low as possible while maintaining an appropriate level of safety, there was an inevitable progression toward economy in materials and structure, and a consequent lowering of the safety factors built into a given bridge. Each success— each design hypothesis tested and verified, though never proved absolutely—brought further efforts to achieve greater economy.

What distinguished the design of the Tacoma Narrows bridge from such illustrious and magnificent predecessors as the Brooklyn Bridge and the George Washington Bridge was the extreme flexibility of its narrow, thin, two-lane roadbed. Unfortunately, this elegant, streamlined design gave the bridge a pronounced tendency to oscillate vertically under windy conditions. Even before its completion, workers had felt seasick as a result of its motion. Later, thrill-seeking motorists would come just for the novelty of driving across "Galloping Gertie's" undulating surface.

Engineers tried to correct the problem but failed. Then, early in the morning of November 7, 1940, with a wind blowing at roughly forty miles per hour, the undulations became more serious. Officials closed the bridge at 10 A.M.—just before it began twisting itself to pieces.

Civil engineers in the United States responded to the Tacoma Narrows disaster by stiffening existing bridges and building new bridges heavy and rigid enough to resist wind-induced motion. In fact, the structure that replaced the original Tacoma Narrows bridge has a much heavier, four-lane roadway. Such bridges naturally flex very little.

It's possible, however, that U.S. engineers learned the wrong lesson from this failure. Simply building heavy, rigid bridges represented a clumsy, obvious solution to the problem. Engineers in Europe generally chose a different approach, conducting a more sophisticated analysis, which showed them how to build elegant, lighter, and probably less costly suspension bridges.

"Had the suspension bridge designers of the 1930s known the history of their genre, and if they had viewed successful design first and foremost as the obviation of failure, the Tacoma Narrows collapse need never have occurred," Petroski claims. "The technical state of the art was simply not, never is, and probably never will be sufficient to guarantee the success of its designs. For success depends upon a constant awareness of all possible failure modes, and whenever a designer is either ignorant of, uninterested in, or disinclined to think in terms of failure he can inadvertently invite it."

Ironically, sometimes it's the most innovative technologies and designs that succeed. The U.S. effort to land a man on the moon was an extremely complicated project, and engineers went to great lengths and took particular care to make sure that every possible detail was handled properly or with a sufficiently large margin of error that no failure would have to be disastrous. The risks were high, but the mission succeeded. The program even survived the explosion aboard a moon-bound *Apollo 13*, a clear example of a catastrophic failure that close attention to reliability overcame.

But after a string of successes, it's easy to become overconfident—even with an intrinsically risky technology such as space travel. Overconfidence led to the tragic loss of the space shuttle *Challenger* in 1986.

Nowadays, engineers may suffer from overconfidence in their ability to analyze systems—to use elaborate computer programs to test mathematical models of structures to "prove" that proposed engineering designs will work. "This is a misguided way to 'prove' a design," Petroski contends. Engineers don't always distinguish carefully between the mathematical version of the structure that the computer sees and the one that will actually be constructed out of real materials by real workers and then used by real people. Mathematical models don't always capture the factors that prove crucial in how well a structure functions.

Moreover, making minor changes in a design can cause big problems. Too often, engineers decide that a certain change is unlikely to have a large effect and neglect to reanalyze the modified design to make sure there are no deleterious, large-scale side effects.

It was just such a minor change in a design that caused the collapse of skywalks in the Kansas City Hyatt Regency Hotel in 1981, where 114 people died and 200 more were injured. Somewhere along the way, someone decided to replace a single long rod, which supported two walkways, by two shorter rods offset from one another to perform the same function. The original design was deemed too difficult, if not impossible, to install. But because the original design itself had virtually no margin of safety, the slightly weaker support structure in the modified design proved unequal to its task.

Such a situation is well known to computer programmers; changing one line of code or inadvertently leaving out or mistyping a single character can have dreadful consequences, just as a single link in a DNA chain can wreak havoc in the human body. Yet it's not unusual for programmers to do the same thing the skywalk designers did when they deviated from the specifications to simplify the task, figuring that it could still meet the intent of the requirements. A similar situation occurs when software is modified. "The minute you make a change in a piece of software, you don't know if you have made it better or worse," David Parnas says. It's hard to escape having to test and analyze the software all over again.

Petroski's call for increasing the emphasis on engineering history and case studies in the education of engineers also applies to software engineering, even though the field itself is extremely young compared with the more established branches of engineering. The basic principles of design, and the importance of learning from failure, apply to all types of engineering.

Unfortunately, the case-study literature for disasters involving computers and software is sparse, despite the Risks forum and Peter Neumann's efforts to get experts to write in detail about their experiences with failures. Even the Risks forum relies too much on reports in newspapers and other media to be sufficiently precise and authoritative for advancing software engineering practice. But a handful of cases have gained celebrity status, partly because they are among the few that have been documented in considerable detail. One is the space shuttle synchronization problem—the "bug heard 'round the world" (see Chapter 1). Another is the analysis by Nancy Leveson and Clark Turner of the Therac-25 incidents (see Chapter 2). A third is the failure of AT&T's telephone network in 1990, which vividly demonstrated how even the most reliable and scrupulously tested systems can fail.

**TODAY'S LONG-DISTANCE TELEPHONE** networks are technological marvels. Telephones serve as the nerve endings of a vast, intricate electrical web, while the electronic switching station acts as the system's brain. And there generally isn't just one brain. In AT&T's network, more than 115 switching stations share the know-how needed to complete a call from one local company's central switching station to another. Together, AT&T's system handles about

115 million calls within the United States and roughly 1.6 million overseas calls per day.

Several decades ago, humans played the central role in switching systems. An operator, almost invariably female, sat in front of a vertical panel punctured by thousands of metal-rimmed holes, each one equipped with its own little lightbulb and labeled with a neatly printed number. When a bulb lit up, the operator knew that the handset of the telephone at the other end of the line had been taken off the hook. For billing purposes, she routinely recorded the number of that phone in a logbook. The operator then plugged in her answering cord and announced her presence by saying "operator" or "number please." The customer then told her the number to be reached, and the operator wrote it down. If the number was on the switchboard, the operator pulled out a long elastic cord from a shelf at the bottom of the switchboard and slipped just its tip into the appropriate hole. If she heard that the line was busy, she pulled out the cord and plugged it into another hole that gave a busy tone. Otherwise, the operator pushed the cord's jack all the way into the hole, a relay was tripped, and the telephone of the person being called rang. If someone took the handset off the hook, a conversation could begin. The operator unplugged her answering cord and, while handling other calls, waited until the lightbulb went out, indicating the call was over and the caller's handset was back on the hook. She then unplugged the cord, which would zip back into place, ready for the next connection. If the call was long-distance and couldn't be handled by her own switchboard, the operator plugged the call into a trunk line, which led to a switchboard higher up in the hierarchy, where another operator waited to pass on the call. Sometimes, making a long-distance call required going through a long chain of operators.

Modern, automated switchboards do the same kind of job more quickly and economically. Each of AT&T's 115 or so long-distance network switches is a maze of electronic equipment housed in a large room. Known as the 4ESS, each switch is really a gigantic, special-purpose computer that not only connects calls but also handles billing and a variety of other telephone services. At its heart is a set of software-driven processors that controls the entire switching system.

Unlike older electromechanical systems that relied on relays and connections over physically separate wires, a modern electronic system allows many calls to be switched over a single physical path. In an exquisitely timed juggling act, the circuitry slices up incoming calls into thousands of parcels a second, labels the samples, then interlaces different calls to pass them along on a single path, like beads on a string. Although customers are connected only intermittently, the quality of voice transmission suffers no noticeable degradation.

Each electronic switch is connected to every other switch, and software orchestrates the entire operation. The computer program for each 4ESS switch runs to more than four million lines of code, written largely in a proprietary computer language. About half of this software is devoted to recovering from faults. It ensures that switches can safely take themselves out of service after detecting a problem, check the problem out, then reset themselves, usually without any callers becoming aware of the interruption.

It takes more than 150 engineers and technicians to keep 4ESS software up to date with new features and improvements. Four or more times a year, they make changes in various modules of the program for each switch. Once a year, the system goes off for about four minutes in the middle of the night for a complete overhaul. In general, changes are introduced cautiously. Sometimes as many as three versions of the software are in use at the same time, as new features are tried out at a few test sites before their nationwide introduction.

All of this effort is aimed at achieving a high level of reliability and speed in connecting calls. Altogether, fewer than one in a billion calls fails to get through. Achieving such a high level of performance has required a tremendous effort, especially in making sure that the software functions properly.

On Monday, January 15, 1990, AT&T's first indication that something might be amiss appeared on a giant video display at the company's network control center in Bedminster, New Jersey. At 2:25 P.M., network managers saw an alarming increase in the number of red warning signals appearing on many of the seventy-five video screens that show the status of various parts of AT&T's worldwide network. The warnings signaled a serious collapse in the network's ability to complete calls within the United States.

Normally, when a local exchange delivers a telephone call to the network, it arrives at one of the 4ESS switching centers, each of which can handle up to 700,000 calls an hour. The switch immediately springs into action. It scans a list of fourteen different routes it can use to complete the call—a direct route from switch to switch or thirteen alternative paths that pass through an intermediary switch. At the same time, it hands off the telephone number to a parallel, signaling network invisible to the caller. This private data network allows computers to scout the possible routes and determine whether the switch at the other end can deliver the call to the local company it serves.

If the answer is no, the call is stopped at the original switch to keep it from tying up a line, and the caller gets a busy signal. If the answer is yes, a signaling-network computer makes a reservation at the destination switch and orders the original switch to pass along the waiting call. The call is passed along after the destination switch makes a final check to ensure that the chosen line is functioning properly.

The whole process of passing a call down the network typically takes four to six seconds. Because the switches must keep in constant touch with the signaling network and its computers, each switch has a subsidiary computer program that handles all the necessary communications between the switch and the signaling network.

It was a defect somewhere in this system that triggered the 1990 network failure. To bring the network back up to speed, AT&T engineers first tried a number of standard procedures that had worked successfully in the past. This time, the methods failed. The engineers realized they had a problem they had never seen before. Nonetheless, within a few hours, they managed to stabilize the network by temporarily cutting back on the number of messages moving through the signaling network. They cleared the last defective link at 11:30 that night.

Meanwhile, a team of more than a hundred telephone technicians tried frantically to track down the fault. By monitoring patterns in the constant stream of messages reaching the control center from the switches and the signaling network, they searched for clues to the cause of the network's surprising behavior. Because the problem involved the signaling network and seemed to bounce from one switch to another, they zeroed in on the so-called System 7 signaling software that permitted each switch to communicate with the signaling-network computers.

The day after the breakdown, AT&T personnel removed the apparently faulty software from each switch, temporarily replacing it with an earlier version of the communications program. A close examination of the flawed software turned up a single error in one line of the program. Just one month earlier, network technicians had changed the software to speed the processing of certain messages, and the change had inadvertently introduced a flaw into the system—a problem that resided in a handful of lines of confusing, excessively complicated code.

Upon finding the error, AT&T engineers could begin to reconstruct what had happened. They discovered that the incident had started when an AT&T switching center in New York City, in the course of checking itself, found it was nearing its limits and needed to reset—a routine maintenance operation that takes only four to six seconds. The New York switch sent a message via the signaling network, notifying the remaining switches in operation at the time that it was temporarily dropping out and would take no more telephone calls until further notice. When it had finished resetting, the New York switch signaled to all the other switches that it was open for business by starting to distribute the calls that had piled up during the brief interval when it was out of service.

One switch in another part of the country received its first message that a call from New York was on its way, and started to update its information on the status of the New York switch. It made a note of the fact that this switch could now accept incoming calls again. But in the midst of this operation, it received a second message from the New York switch, which arrived less than a hundredth of a second after the first.

That's where the fatal software flaw surfaced. Because the switch's communications software was not yet finished with the information from the first call, it had to shunt the second message aside. Because of the programming error, the switch's processor mistakenly dumped the data from the second message into a section of its memory that was already storing information crucial for the functioning of the communications link. At this point, all was not yet lost because the system was designed to recover from such faults. The switch detected the damage and promptly activated a duplicate, backup link, allowing time for the original communication link to reset itself.

But it was a busy time, and the backlog from the New York switch was still coming through. Unfortunately, another pair of closely spaced calls put the backup processor out of commission, and the entire switch shut down temporarily. These delays caused further telephone-call backups, and because eighty of the switches had the same software containing the same error, the effect cascaded through the system. The instability in the network persisted because of the random nature of the failures and the constant pressure of the traffic load within the network.

Of the roughly hundred million telephone calls placed with AT&T during the nine hours that it took to stabilize the system, only about half got through. The breakdown cost the company more than $60 million in lost revenues and caused considerable inconvenience and irritation for customers. Despite carefully honed techniques for building software, extensive testing, and a system designed to facilitate recovery from a failure, a parcel of flawed code had managed to slip through to inflict harm to the network. Although the software changes introduced the month before had been rigorously tested in the laboratory, no one had anticipated the precise combination and pace of events that would lead to the network's near collapse. The fact that some of switches were not using the flawed software allowed parts of the network to operate while AT&T personnel were tracking down the problem.

"Ironically, the new software was designed to improve the signaling performance of the AT&T network," Karl Martersteck, AT&T's vice president for network development, remarked two days later.

"Software development is a fascinating challenge," said Martersteck. "We design and produce very high quality software. But we are painfully aware that quality cannot be tested in, because laboratory testing is just an emulation of the real world and there is no way to guarantee that every possible contingency is tested.

"Why didn't the problem surface during load testing? The odds against a processor receiving too many requests within 0.01 second are so small that even our rigorous testing didn't stress the software sufficiently. It's a tyranny of numbers problem: There is an astronomical number of possibilities to consider."

In their public report, members of the team from AT&T Bell Laboratories who investigated this humbling incident stated: "We believe the software design, development, and testing processes we used are based on solid, quality foundations. All future releases of software will continue to be rigorously tested. We will use the experience we've gained through this problem to further improve our procedures."

The key problem stemmed from the fact that AT&T's 4ESS switches don't stand alone but are connected into a network. Such connectivity allows for "network bugs." In other words, even though the switch hardware and software may function properly, faults may lie in the way things are connected or in the way those connections could help the consequences of an error to spread. Thus, it's insufficient to test the software on one switch; several switches must be linked, and the software must be run on the resulting combination.

At the time of the breakdown, AT&T already had in place its so-called integrated test network (ITN) for subjecting software to conditions as close as possible to the real world. Serving as a microcosm of the AT&T telecommunications system but completely separate from it, the ITN has the same switching, database, and signaling capabilities as the real network. It's here that software comes after initial testing at Bell Laboratories. But the ITN still wasn't able to pick up the bug responsible for the January network disaster.

The company also had software—in the form of an expert system—that automatically monitors the circuits in all 4ESS switches. Called TERESA (trouble evaluation and resolution via expert systems applications), this elaborate computer program consists of a web of rules that encodes a large amount of specialized knowledge about how the network should operate. Acting much like a human expert, such a system can automatically detect switch problems, isolate the trouble, and inform technicians of the type and location of the faulty equipment.

TERESA comes into play when the software installed in the 4ESS switches notices something abnormal in a particular telephone circuit required to complete a call. While the switch searches for another, clear circuit to use, it sends out an alarm in the form of an electronic message dubbed a "trouble ticket." This message provides information about the potential fault and background data on the particular circuit and trunk line with the problem. TERESA automatically picks up the trouble ticket and analyzes the situation. More often than not, it finds no real problem and instructs the switch to return this telephone circuit to service. If something proves faulty, it runs various tests and checks both ends of the circuit, working toward the middle. About 10 percent of the time, it fails to pinpoint the difficulty and leaves a message for a human technician.

By the end of 1989, seventeen TERESA systems were operating in various centers throughout the network. Like their human counterparts, these "experts" communicated with one another to help solve problems. Because they could do it twenty-four hours a day, seven days a week, it was almost like having a top expert technician on duty at all times in every switching office.

But TERESA could provide little assistance during the January 1990 outage. There were too many alarms and too much information flowing into the network emergency service center for it to handle. Only human experts could effectively track and analyze the flood of data, and it took them nine hours to bring the situation under control.

"TERESA is the foundation of a new system we're working on that will automatically analyze equipment alarms and data from the Switching Center Control System and present those centers with processed information," says AT&T's Terry Miller, who oversaw the development of TERESA. "We have hopes that systems like this could significantly cut down on the time needed to respond to those kinds of events on the network."

THE DISRUPTION of service in January 1990 and several subsequent outages on a smaller scale prompted a considerable amount of soul-searching and concern at AT&T and at other telecommunications companies, which face similar vulnerabilities (see Chapter 1). It prompted new efforts, including the formation of a special group within AT&T Bell Laboratories, aimed, in some sense, at regaining control of these systems.

The systems are now so complex that no individual can possibly have a sense of what's going on. "Problems have arisen because this sense of where you are [in the system] has been lost," says Michael Meyers of AT&T Bell Labs in Naperville, Illinois. "This is a new class of problems for us," he notes. "Before [January 1990], we would have said it could not have happened."

In general, one can tell how reliable a system is by how it fails. Intrinsically unreliable systems fail in simple ways that are relatively easy to pinpoint and understand, as in the cases of the *Phobos* spacecrafts or the Therac-25 radiation machines. A reliable system is built so that it can't fail in a simple way. On the rare occasions when a failure occurs, it nearly always has to be complicated and hard to find and understand, as it was in the AT&T case.

The software engineers at AT&T are continuing to restructure their four million lines of code to make it easier to understand and test, which also means that the program must be rigorously tested again and again. Just as in the case of the nuclear reactor emergency shutdown software (see Chapter 3), it's a matter of engineering judgment to decide how much rechecking to do for any given change in the program. A full revalidation of the whole program can't be performed after every minor alteration. It's simply not cost effective.

At the same time, the marketplace adds to the difficulty of maintaining a telecommunication network's reliability. AT&T's equipment and software must interact with signals from the networks of local companies, and the local companies work with a number of long-distance carriers, along with cellular phone systems and other services. In such a polyglot conglomeration, all kinds of faults are possible. An ailing element in one part of the system can infect the rest, unless precautions are built in to prevent or mitigate the transmission of such potentially harmful vectors.

In response to mounting concerns about the reliability of the nation's phone system, the Federal Communications Commission held a closed meeting in September 1991, bringing together representatives of the seven regional Bell operating companies, long-distance carriers, equipment manufacturers, state regulatory agencies, and users. One result was an agreement by Bell Communications Research (Bellcore) to draft a proposal on network testing and interconnection as a step toward standards for ensuring reliable connection from one caller to another, anywhere in the system. There was also agreement that the phone company service providers and equipment vendors needed to plan for better sharing of information.

Nearly as serious are the problems created by the addition of new services to both local and long-distance networks. Allegedly offered in response to customer demand but aimed largely at keeping ahead of the competition, services such as call-waiting, call-forwarding, call-screening, call-blocking, speed-dialing, and more than a dozen others add up to huge headaches for the programmers who must incorporate them without hurting the rest of the system.

The trouble is that one feature may interfere with the operation of another and put it out of commission or produce some other peculiar, unexpected result. Solving the problem often requires some creative rewriting of the software—and increased complexity. For example, in a normal two-person telephone call, either party can end the call by hanging up. On the other hand, a call to the emergency number 911 can be disconnected only by the 911 operator. This becomes a problem if the customer calling 911 also has the call-waiting feature and a phone call arrives. The person calling 911 could conceivably put the 911 operator on hold to answer the new call. But this means that the 911 operator no longer controls the call, which violates the accepted rules of operation for an emergency service. Such a scenario can also tie up resources that might otherwise be used for responding to other life-threatening situations.

Other features such as call-forwarding and multiparty calling add to the confusion. In the past, each telephone number identified a single destination and a single user. Call-forwarding alters this association because the number called is not that of the telephone that rings. Therefore, current specifications must deal with situations that did not arise before. For example, the owner of a telephone may screen out calls from selected callers. Should the owner be allowed to screen out a forwarded call intended for a different person? Another problem arises when the call-forwarding user has call-waiting but the destination number does not. Should the call-waiting tone sound on the destination phone?

And it gets worse! Suppose Andy is speaking to Betty and receives a call-waiting tone from the switching center. Andy can simply depress the hook for an instant to accept the new call. Similarly, if Andy is speaking to Betty and wants to add Charlie via the three-way calling feature, Andy also briefly depresses the hook. However, if just as Andy depresses the hook to activate three-way calling, an incoming call from Dave arrives at the local switch and a call-waiting signal is sent to Andy, the switch may interpret the momentarily depressed hook as a signal to put Betty on hold and set up a call between Andy and Dave.

This turns into a serious timing problem because even the small delays that inevitably occur in the transmission of signals between telephones and switching centers can cause trouble. The problem arises because each telephone and switch acts independently of the others, and due to propagation delays, the incoming and outgoing signals are received, interpreted, and sent in different orders. As a result, when Andy expects to activate three-way calling, the switching center may put call-waiting into effect.

Every new feature added to the telephone system can change

the behavior of features that already exist, and perhaps even "break" them, crashing the entire system. "Feature interaction is one of the most difficult problems we face," AT&T's Michael Meyers says. "But I can't think of a case where we've excluded a feature. We've always figured out a way to do it."

To identify and mitigate potential interactions, telecommunications specialists are studying ways of describing or specifying exactly what current and proposed features really do. But progress in developing such specifications has been slow.

"Perhaps the most difficult problem of all is the incoherent, unprincipled, and ad hoc behavior of current telecommunications systems, which cannot be specified compactly and comprehensibly, regardless of the quality of the available specification technology," says Pamela Zave, a computer scientist at AT&T Bell Laboratories, who specializes in techniques for specifying what computer programs do. She adds, "Dreadful as these problems of evolution might be, the cost of inaction might be greater. The problem of feature interaction is getting rapidly worse because of pressures for faster feature development, technological diversity, decentralization of function, and pressures against monopolies at all levels of telecommunications service."

In other words, there is a real danger of drifting into a telecommunications Babel.

Aside from the problem of special features, the phone system also has to deal with congestion when a telecommunications network suddenly becomes overloaded. AT&T and its competitors already have systems in place that allow them to turn off certain services or reduce the number of calls handled in a given time period. They can require more time to make connections during peak periods or allow more busy signals to occur. But congestion control is actually a lot more difficult than it sounds. One researcher has labeled it the most important but least understood issue in real-time communication networks.

When congestion occurs, the tasks of the network management

system are to minimize the congested area, to shorten the duration of congestion, and to avoid escalation of disruptions. The principles for handling these tasks are not so different from what a human telephone operator once did. Namely, traffic is prevented from entering the congested area, calls to and from unaffected areas are rerouted around the congested area, and information is provided to the callers to cut back redialing. But there are numerous potential traps in automated congestion control. For example, network management messages conveying information about an outage can flood the system, adding to the traffic and preventing repair messages from getting through, which in turn triggers more management messages, and so on.

To avoid faults in the first place, companies like AT&T have placed a great deal of confidence in automated methods of testing and verifying software. Such systems are supposed to evolve gradually toward a self-healing or self-stabilizing state. But who's guarding the guards? Are the automated methods—computer programs themselves—reliable enough for their complex tasks? At the same time, old formulas for software reliability, applicable to single programs on one computer, are no longer effective enough to capture the behavior of a network that's perpetually in a state of evolution and continually being modified on the fly.

"It is quite evident that the current technology is far from being able to deal with the serious obstacles on the road to reliable software and communication," says Fan Chung, a mathematician who worked at Bell Communications Research before moving to the University of Pennsylvania. "We usually hear about the lack of automated tools for testing and verification, but the truth is the lack of knowledge. We can imagine the analogous situation of trying to solve problems in quantum chemistry by applying the 'art' of alchemy. What we really need at present is a sound scientific foundation upon which the advanced technology in broadband communication and multimedia networks can be built."

Adding a further realm of complexity, computer personnel, technical workers, users, and others who interact with the system arc far from error free. A backhoe operator can accidentally cut through a bundle of optical fibers and sever a crucial, high-capacity communications link. Maintenance workers may neglect to check that batteries are kept charged to scrvc as emergency power supplies. A programmer, sleepy from a long day and night spent fixing a problem, could inadvertently write an instruction in a form that shouldn't be used. A user may push the "wrong" button.

**THE TYPICAL RESPONSE** to such problematic situations has been to take the human out of the loop as much as possible by introducing automation. But such a strategy simply shifts responsibility to a different group of humans—to those charged with designing, building, and programming the automated systems.

One can argue that automation—whether in a telecommunications network, a medical device, an avionics system, or a power plant controller—makes the system more reliable than the human-centered system it replaced. That's probably true when the human-centered system is replaced by an equivalent automated one, but usually the automated system ends up being far more complicated than what it replaces. The additional margin of safety that automation was to provide often disappears in a new layer of poorly understood complexity. Just as computers make it possible to complete long-distance telephone calls with greater rapidity than ever before, they also set the stage for offering new services, such as call-waiting, which add new troubles.

There arc risks in all of the technologies that now permeate our lives. Electricity and natural gas are dangerous, and it would certainly be possible to write books about electrical systems and how they fail or about natural gas systems and how they fail. But we have learned to live with these potential hazards and to accept the risks, though occasionally with some trepidation, as in the case of underground gas lines in earthquake fault zones or ships transporting huge cargoes of liquefied natural gas.

In the case of software, however, the changes in the technology are so rapid and the scope of its domain is so vast and still increasing that learning has to take place on the fly. It's a precarious situation in which the risks are not yet clearly defined. Computer professionals joke that if the automobile industry were like the computer industry over the past thirty years, a Rolls-Royce would cost $5.00, would get three hundred miles to the gallon, would cross the country in a matter of minutes, and once a year, would explode, killing all passengers!

Not every backyard sports a tanker carrying liquefied natural gas. But computers are ubiquitous in daily life. A typical individual may encounter computers or programmed microprocessors in kitchen appliances, heating and air-conditioning systems, television sets and videocassette recorders, and even light switches. New automobiles carry an array of microprocessors, controlled by as many as thirty thousand lines of code. Computers synchronize traffic lights and operate rapid-transit lines. Using "fuzzy logic" and other schemes, they even schedule elevators to cut down waiting times.

Airlines, travel agencies, banks, retailers, and innumerable other businesses now rely on computers. Supermarket checkout rests on the vagaries of prices stored in computer databases. Restaurants use computers for managing food supplies, billing customers, maintaining menus, posting specials, and even paging waiting guests. The vast communications web, from desktop systems to cellular phones, that ties our society together depends on computers and software.

And there is much more to come. There is no escape.

A decisive choice confronts us. If we can gradually rein in the digital beast that has so pervasively and so rapidly come to dominate our lives, we will learn how to minimize the risks that flawed computer systems and misguided automation pose to society. The alternative is to continue our headlong plunge into computer-driven automation, allowing the virtual free-for-all to continue with its attendant triumphs and risks.

Spending the time and effort required to learn how to catch mistakes reflects a recognition that no human-designed system is perfect. We must also acknowledge that completely guaranteed results are beyond reach, and we must act accordingly by entrusting computers only with what we can afford to lose.

"The problem is intrinsically unsolvable, but you can always do better," Peter Neumann says. "It's a question of system design, of experience, of good software engineering techniques, or recognizing risks, and of continually adapting to a changing environment. There are no easy answers."

Computer programming is a highly creative, complex effort, requiring many of the same skills that go into writing a good novel. Every good program is unique. There's no single, correct approach, no single, ideal computer language. Programming requires an exacting blend of ingenuity, imagination, discipline, and skill. Imagine a novel with the length and complexity of an encyclopedia, and you have an idea of what today's software projects often involve.

Computer scientist Frederick P. Brooks, Jr., of the University of North Carolina, has pondered the state of the art of computer programming and software engineering. "Of all the monsters who fill the nightmares of our folklore, none terrify more than were wolves, because they transform unexpectedly from the familiar into horrors. For these, one seeks bullets of silver that can magically lay them to rest," he wrote in a famous 1986 essay. "The familiar software project has something of this character . . . , usually innocent and straightforward, but capable of becoming a monster of missed schedules, blown budgets, and flawed products. So we hear desperate cries for a silver bullet, something to make software costs drop as rapidly as hardware costs do."

He went on, "But as we look to the horizon of a decade hence, we see no silver bullet. There is no single development, in either technology or management technique, which by itself promises even one order of magnitude improvement in productivity, in reliability, in simplicity."

Simply put, software design is hard work, and nothing will fundamentally change that. The computer offers infinite flexibility; the same machine is capable of doing so many things. But capturing in the bare bones of the lines of a computer program what the human mind can imagine remains the central challenge, subject to the fallibilities of the human mind at work.

"There is inherently no silver bullet," Brooks says. "Not only are there no silver bullets now in view, the very nature of software makes it unlikely that there will be any—no inventions that will do for software productivity, reliability, and simplicity what electronics, transistors, and large-scale integration did for computer hardware."

He emphasizes, "There is no royal road, but there is a road."

The safety and trustworthiness of a safety-critical system rests on a tripod made up of testing, mathematical review, and certification of personnel and process. Unfortunately, anyone can assume the title of programmer, or even software engineer. There is no authority in software engineering comparable to those that certify professional engineers in other areas. Writing software for important applications requires more discipline than many programmers appear to apply, David Parnas contends. No other essential technology, from the building trades to automobile design and manufacture, remains as unregulated as software design.

"A responsible software designer would make it a point to know and abide by the standards and practices of software engineering; and the designer would forever be a student of the breakdowns of software systems and would where necessary work to alter the standards of the field," says Peter Denning, a computer scientist at George Mason University in Fairfax, Virginia.

However, there is strong resistance within the freewheeling computer community to the notion of certification. It would stifle innovation, some argue. Others contend that it would create the modern version of a priesthood, with only a select few having the privilege of using authorized techniques to work on certain systems.

Some also argue that fears of liability are enough to weed out poor performers. Yet, letting the courts decide has proved an inefficient, unreliable mechanism for achieving quality and promoting reform. Moreover, fear of legal action has the unfortunate by-product of encouraging both individuals and companies to hide their mistakes and slovenly practices, and the lessons that can be learned from defects and failures are lost.

Neumann describes licensing of software professionals as a double-edged sword. Any professionally acceptable certification process would typically define a minimum set of skills that professional programmers or software engineers ought to have. But the qualities required of programmers dealing with life-critical systems must go far beyond the minimum. They must have a tremendous amount of experience, creativity, imagination, along with a sense of what won't work and a conservative attitude toward software development, Neumann says. No certification process could ever ferret out those traits.

Nonetheless, if we allow the technological anarchy still evident in the computer enterprise to push us into a bewildering digital netherworld, then we could lose all chance of making meaningful choices. It's not difficult to imagine a society in the not-too-distant future fatally enmeshed in a tragically flawed technological creation that no one really understands. We would face mysterious, all-powerful entities whose whims dictate our daily lives in ways that we cannot fathom—a creation that apparently works perfectly most of the time yet sometimes inexplicably and unexpectedly displays bizarre, even destructive behavior.

Unlike a detective novel in which the final pages unmask the murderer and neatly tie up loose ends, the software story does not have a satisfying ending. The bug hunters face a daunting but crucial task. Their success in dealing with defects will determine the degree of digital uncertainty with which we will all have to learn to live. On one side, fuzzy thinking and increasing complexity are leading to unwieldy, defective computer systems. On the other side, increasingly strident consumer complaints, pressures to improve productivity within the computer industry, and legal threats against faulty systems are stimulating improvements in quality. Where the balance will settle remains in question.

Our society depends on the reliable—if not always correct—functioning of computers. More than anything else, the computer is an amplifier of our mental prowess. It amplifies our genius and our flaws, extrapolating them at lightning speed to what may be unreasonable extremes. If we don't clean up our mental act, if we fail to think clearly and intelligently about the misconceptions and mistakes built into computer programs and systems, we may find those systems punishing us for our sloppiness. That's what has already started to happen.

Balanced against this dark vision is the irresistible, deeply seductive promise of computation. It offers us a chance to play with our thoughts and our dreams in ways that nothing else in the world allows. We put up with the clumsiness, the bugs, and the irritations because the promise, fueled by our imaginations, is so strong. We have but to express our desires, and it's possible to imagine the chameleonlike computer sating those varied longings.

Software, a vehicle of dreams and aspirations, exists perpetually in the distance. It's never finished, it never quite catches up with our visions, and we are relentlessly drawn forward into an uncertain future.

# Perpetual Glitch

**DEFECTS IN COMPUTER SYSTEMS** are nothing new. Indeed, failures litter the history of computers. Even use of the term "bug" to signify a computer problem goes back to the earliest days of automated computation.

One of these pioneering machines of the 1940s was the Mark II. Extremely primitive compared with today's digital wonders, this huge contraption relied on electromechanical relays instead of electronic circuitry to perform calculations. These relays were nothing more than metal switches that could be forced back and forth between two positions by means of electrically controlled magnets. Installed in a vast space resembling an airplane hangar at the Naval Proving Ground in Dahlgren, Virginia, this clicking and clanking behemoth was used by Navy personnel to compute ballistics tables and solve basic mathematical equations.

One day the Mark II suddenly stopped, and Navy staff members established the cause as a failed relay. They opened up the machine to take a close look at the faulty device. Here is the way Grace Murray Hopper, a programmer working with the Mark II, described what they found:

> Inside the relay—and these were large relays—was a moth that
> had been beaten to death by the relay. . . . Now, Commander

Howard Aiken had a habit of coming into the room and saying, "Are you making any numbers?" We had to have an excuse when we weren't making any numbers. From then on if
we weren't making any numbers, we told him that we were *debugging* the computer.

The term seemed just right, neatly capturing the maddening frustrations of the never-ending pursuit of digital perfection.

This was also just the beginning. Every day sees new additions
to the catalog of computer faults and failures. For example, in the
spring of 1995, some people were shocked to find that certain
income tax preparation programs generated inaccurate or incomplete returns. In another widely reported incident, an expert witness at the trial of O. J. Simpson made a mistake in a computer
program and calculated incorrect probabilities involving DNA
evidence.

In June 1995, the national heavy-ion accelerator in Caen, France,
was shut down following the discovery of a failure in a safety system designed to prevent staff from entering a cyclotron while it is
in operation and thus being exposed to ionizing radiation. The
problem was eventually traced to an error in software that had been
recently modified.

Other reports highlighted sporadic failures of air traffic control
computers in the United States and elsewhere, reliability problems
(including an excessive rate of false alarms) in a wide range of computer-based patient monitors used in hospitals, a variety of glitches
involving computers handling financial transactions, and serious
voting machine breakdowns during the November 1995 elections
in Pennsylvania.

People in the computer industry are well aware of the innumerable difficulties, but they sometimes defend their failed practices as part of the price we must pay for progress and innovation.
One often hears the statement "There are always bugs." Such an
outlook makes it sound as if defective products and flawed sys-

tems are inevitable. And it quickly becomes an excuse for shoddy work—software products that are badly designed, poorly tested, and rushed into the marketplace.

Computer programmers and system designers can do considerably better—even though they can never guarantee perfection. The key problem they have to cope with is burgeoning complexity. Computer programs have evolved into monsters having millions of instructions, incorporating all kinds of intricate operations and interactions. And the irresistible trend is toward even greater complexity.

In the computer realm, we can envision all kinds of possibilities, and, most of the time, there's nothing to stop us from trying to realize these goals, whether in virtual reality, electronic commerce, the digital library, the "smart" home, or the information superhighway. No one sells new products that have fewer features and less glitz than offered by their predecessors or competitors.

Complexity also arises in another way. In many situations, computers help entrench and extend ill-advised practices, often taking away the incentive to simplify, rationalize, or replace cumbersome procedures.

For example, border police monitoring the entry of people from Mexico into the United States must fill out as many as 51 forms for each arrest they make. It can take hours to complete these forms, using up time that could be used for enforcement. However, instead of looking for ways to simplify the system and reduce the paperwork, the Justice Department's solution was to bring in computers to speed up the old process.

We're addicted to complexity and what it appears to offer in both commercial and custom software, and our addiction leads to computer systems that outpace the ability of even top experts to analyze, test, and understand them. As a result, the road ahead has many more potholes, detours, and dead ends than marketers of the information age generally like to admit.

In 1995, a group of computer professionals established the National Software Council (NSC) to focus on software issues that affect the economy and security of the United States. Its founders wanted a forum in which members could discuss, formulate, and recommend policy initiatives relevant to such matters as software quality, research, and education and training.

The council highlighted a number of concerns:
- The United States continues to have major problems developing large and complex software systems. Projects often suffer huge cost overruns, produce defective and poor-quality systems, and end up causing economic loss.
- No university in the United States offers a software engineering specialty at the undergraduate degree level, severely limiting the supply of qualified, entry-level software engineers available in the job market.
- Software companies generally lack the understanding and expertise needed to plan for and implement software quality programs.
- There exists no broadly based effort to collect and analyze data that would allow businesses and government to evaluate the impact and value of software investments as part of decision-making processes.

The council's inaugural event—a "National Software Summit"—in November 1995 attracted about ninety professionals from industry, academia, and government. The participants generally agreed on the need for such a forum to share recent developments, to explore the possibilities of fruitful collaborations, and to help educate policy makers and the general public concerning software's vital role—and its risks—in today's society.

In a keynote address, Lawrence Bernstein of AT&T Network Systems emphasized the need to change the way people think about software. "We have an intense and important focus on software processes [specifying how software is developed], but we lack a corresponding focus on software product performance," he argued. "We must move to a risk management approach.

"Moreover, safety-critical software technology cries out for trained and skillful practitioners," he continued. "Government licensing of software professionals practicing in this software field is long overdue." But how to do it effectively remains highly controversial.

The NSC represents an important initiative for the computer community, but it's not yet clear whether it will develop into an effective organization. Meanwhile, computer technology advances, new problems arise, and old problems get swept under the rug of faster microprocessors and larger memories.

As computers slip into invisibility and ubiquity—in automobiles, homes, offices, businesses, and innumerable other environments—computer users themselves must take a more active role in making their needs known and their complaints heard. They need to stop blaming themselves or feeling guilty about problems that are really not their fault. They should stop feeling that these difficulties are theirs alone.

As customers and users, we need to know something about how computers work and the limits of their capabilities. Though we can't expect perfection, we should be demanding greater reliability in the systems that directly affect our lives. And we should understand the risks involved in not insisting on improved quality, so that we can make intelligent choices and decisions on how much and what we should entrust to computer systems.

We should be able to ask a bank, an airline, a supermarket, or any other business why its computers happened to malfunction or why an error was made, and we should get a reasonable response. We should be prepared to take our business elsewhere if problems persist or if no one pays attention to our concerns.

At the very least, software publishers and vendors should be required to own up to their mistakes. They should maintain and make available to their customers a complete, up-to-date list of any reported defects. Currently, most companies make changes quietly, slipping corrections into new versions of their software or posting fixes on various electronic bulletin boards.

A significant change in the way the software industry typically handles bugs occurred in October 1995 when the Netscape Communications Corp. announced a program of cash and prizes to anyone finding serious flaws in test versions of its software. By turning bug hunting into a contest, the company admitted that it needed help in rooting out its bugs and showed it takes glitches in its software seriously. Several other companies now offer a variety of incentives to encourage bug reporting.

The consequences of remaining inattentive to potential problems are serious. What we may be drifting into is not so much a time of great disasters involving computers—though they undoubtedly are bound to occur—but a state of perpetual glitch.

In today's office, for example, it's not unusual to have something in the networked computer system not working on any particular day. One time, it's a bug in the electronic mail system. Another day, the computer simply crashes. Occasionally, the virus checker shuts down the network, though it turns out to be a false alarm. On another occasion, the backup procedure fails because of a defect in the supposedly bulletproof backup software. Files get lost; commands sometimes elicit the wrong response or no response at all. Sometimes a glitch mysteriously appears (perhaps on the night of a full moon) and then just as mysteriously disappears when the system is rebooted a few hours later. These are very complex systems.

Programmers and computer users are generally optimists. They tend to believe that the next update, the next operating system, the upgraded software will solve the current problem. But as past experience demonstrates, these fixes and enhancements inevitably introduce new difficulties. It would help if users and programmers more often resisted the siren song of "new and improved" to settle for less glitz and more reliability. Simplicity in the computer world has its virtues.

We've survived the computer age so far not because computers have gotten more capable, faster, and more user friendly. We've survived because human beings are amazingly adaptable. They can learn to put up with anything if they have to, whether it's FORTRAN, Unix, WordPerfect, ATMs, Call Waiting, Windows, or the Internet.

The computer age is really a tribute to our adaptability and resilience. But this learning overhead keeps growing. New computers, new operating systems, new programs come faster and faster. Such an accelerating pace takes its toll. How far can we be pushed before we are all lost in a cyberfog in which things happen to us for no apparent reason?

**Ivars Peterson**
January 1996

# ACKNOWLEDGMENTS

**I HAVE LONG BEEN FASCINATED** by the question of what humans can and cannot achieve and what they can and cannot know. Physical theory and experiment set such limits as the speed of light, and mathematical logic defines what is numerically solvable and what is not. But where the boundaries lie in many situations is often still shrouded in mystery.

A marvellous volume called *The Encyclopedia of Ignorance*, edited by Ronald Duncan and Miranda Weston-Smith and published in 1977 by Pergamon Press, promised "everything you ever wanted to know about the unknown." It set me going along this track. A second book called *No Way: The Nature of the Impossible*, edited by Philip J. Davis and David Park and published in 1987 by W. H. Freeman, furnished important signposts along the way.

I'm grateful to my editors at *Science News*, first Joel Greenberg, then Patrick Young, for allowing me the freedom to explore this particular path. I started as an intern more than fourteen years ago, and my second major article for *Science News* concerned Manuel Blum and his scheme for flipping a coin fairly over the telephone. A year later, as a fulltime staff member, I was writing about William Kahan and the foibles of calculators. I first encountered Victor Basili in 1983 at a conference I attended on computers in military systems. Since then, I have attended many more conferences and lectures, meeting and hearing such individuals as Peter Neumann, Nancy Leveson, David Parnas, and Harlan Mills. Many of them have been the subject of articles in *Science News*. Indeed, a portion of the material in this book has previously appeared, in a somewhat different form, in the magazine.

Undertaking a book of this scope and sensitivity would not have been possible without the cooperation of a large number of people who freely shared with me their ideas, experiences, and insights. I am particularly indebted to Victor Basili, Manuel Blum, Elliot Chikofsky, James Fetzer, William Kahan, Nancy Leveson, Peter Neumann, David Parnas, and Elliot Soloway, who provided a great deal of material and reviewed portions of the manuscript.

I also appreciate the help of Glenn Archinoff, Nancy Blachman, Nelson Blachman, David Chudnovsky, Fan Chung, William Cody, James Cross, Phil Hodge, Bill Kelly, Lisa Kelly, Akhlesh Lakhtakia, Frank McGarry, Peter Mellor, John Osmundsen, Cliff Pickover, Fred Ris, John Shore, Harold Stone, Dick Waters, Dolores Wallace, Peter Weinberger, and Lauren Wiener, who contributed in a variety of ways. My apologies to anyone I have inadvertently omitted from the list.

My wife, Nancy, played a major role in editing the original manuscript. She and my two sons, Eric and Kenneth, were remarkably patient and understanding when I had to ensconce myself in the basement to complete the book. I also wish to thank my agent, Gail Ross, and editors Betsy Rapoport and Richard Gerber Kohl at Random House, who helped bring this book into existence.

# BIBLIOGRAPHY

## GENERAL

Andrews, Edmund L. "The Precarious Growth of the Software Empire." *New York Times*, July 17, 1991.

Chick, Morey J. "Policies and Research Needed for Assessing Risk of Automated Information Used in Human Safety Applications." *Information Management Review* 2 (Summer 1986): 49–62.

———. "Interrelationships of Problematic Components of Safety Related Automated Information Systems." In *Proceedings of the Sixth Annual Conference on Computer Assurance (COMPASS '91)*, June 1991; 53–62. Piscataway, N.J.: IEEE Press.

Computer Science and Technology Board, National Research Council. *The National Challenge in Computer Science and Technology*. Washington, D.C.: National Academy Press, 1988.

———. *Scaling Up: A Research Agenda for Software Engineering*. Washington, D.C.: National Academy Press, 1989.

Davis, Bob. "Costly Bugs: As Complexity Rises, Tiny Flaws in Software Pose a Growing Threat." *Wall Street Journal*, January 28, 1987.

Dunlop, Charles, and Robert Kling, eds. *Computerization and Controversy: Value Conflicts and Social Choices*. Boston, Mass.: Academic Press, 1991.

Eames, Office of Charles and Ray. *A Computer Perspective: Background to the Computer Age*. Cambridge, Mass.: Harvard University Press, 1990.

Enfield, Ronald L. "The Limits of Software Reliability." *Technology Review* (April 1987): 36–43.

Forester, Tom, and Perry Morrison. *Computer Ethics: Cautionary Tales and Ethical Dilemmas in Computing*. Cambridge, Mass.: MIT Press, 1990.

Gibbs, W. Wayt. "Software's Chronic Crisis." *Scientific American* (September 1994): 86–95.

Gleick, James. "Chasing Bugs in the Electronic Village." *New York Times Magazine,* June 14, 1992, 38–40.

Knuth, Donald E. "Algorithms." *Scientific American* (April 1977): 63–80.

Lee, Leonard. *The Days the Phones Stopped: The Computer Crisis—The What and Why of It, and How We Can Beat It.* New York: Donald I. Fine, 1991.

Mellor, Peter. "Can You Count on Computers?" *New Scientist* (11 February 1989): 52–55.

Norman, Adrian R. D. *Computer Insecurity.* London: Chapman and Hall, 1983.

Olson, Steve. "Pathways of Choice." *Mosaic* (July/August 1983): 2–7.

Oman, Paul W., and Ted G. Lewis, eds. *Milestones in Software Evolution.* Los Alamitos, Calif.: IEEE Computer Society Press, 1990.

Ornstein, Severo M., and Lucy A. Suchman. "Reliability and Responsibility." *Abacus* 3 (Fall 1985).

Peterson, Ivars. "Warning: This Software May Be Unsafe." *Science News* 130 (September 13, 1986): 171–73.

Schechter, Bruce. "The Maestro of the Algorithm." *Discover* (September 1984): 74–78.

Simons, G. L. *Viruses, Bugs and Star Wars: The Hazards of Unsafe Computing.* Manchester: NCC Blackwell, 1989.

Thomas, Martyn. "Can Software Be Trusted?" *Physics World* (October 1989): 30–33.

Tomayko, James E. *Computers in Space: Journeys with NASA.* Indianapolis: Alpha Books, 1994.

Walker, Henry M. *The Limits of Computing.* Boston: Jones and Bartlett Publishers, 1994.

Wiener, Lauren Ruth. *Digital Woes: Why We Should Not Depend on Software.* Reading, Mass.: Addison-Wesley, 1993.

Wray, Tony. "The Everday Risks of Playing Safe." *New Scientist* (8 September 1988): 61–65.

## PREFACE: BUG HUNT

Anthes, Gary H. "Why Uncle Sam Can't Compute." *Computerworld* (May 18, 1992).

Brody, Herb. "The Pleasure Machine." *Technology Review* (April 1992): 31–36.

Brooks, Frederick P. *The Mythical Man-Month: Essays on Software Engineering.* Reading, Mass.: Addison-Wesley, 1975.

Dauber, Steven M. "Finding Fault." *Byte* (March 1991): 207–14.

Feder, Barnaby J. "Sophisticated Software Set for Exotic Financial Trades." *New York Times*, March 30, 1994.

Gemignani, Michael. "Whom Do You Call if the Program Doesn't Work." *Abacus* 4 (Spring 1987).

Grimsley, Kirstin Downey. "At the Register, Getting Rung Up . . . and Riled." *Washington Post*, June 8, 1994.

Halpert, Julie Edelson. "Who Will Fix Tomorrow's Cars?" *New York Times*, November 7, 1993.

Hansell, Saul. "The Greedy Cash Machines: 100,000 People Are Overbilled." *New York Times*, February 18, 1994.

———. "Bugs and Squirrels Gnaw Away NASDAQ's Image." *New York Times*, August 3, 1994.

Hecht, Jeff. "Bank Error Not in Your Favor." *New Scientist* (5 March 1994): 17.

Hoffman, Thomas. "Squirrel Crashes Stock Market." *Computerworld* (August 8, 1994): 4.

Kay, Alan. "Computer Software." *Scientific American* (September 1984): 52–59.

Knight, Jerry. "Deposits Vanish in N.Y. ATM Glitch." *Washington Post*, February 18, 1994.

Levy, Steven. "Kay + Hillis." *Wired* (January 1994): 13–17.

McCarroll, Thomas. "No Checks. No Cash. No Fuss?" *Time* (May 9, 1994): 60–62.

Miller, Michael W. "Who Takes Blame When Trades Short-Circuit?" *Wall Street Journal*, November 20, 1990.

Passell, Peter. "Fast Money." *New York Times Magazine*, October 18, 1992.

Pasternack, Andrew. "Big Software Glitch Disrupts Stock Exchange." *Open Systems Today* (March 16, 1992).

Patton, Robert. "Software Skipper." *Scientific American* (November 1993): 108–109.

Peers, Alexandra. " 'Paperless' Wall Street Is Due Next June." *Wall Street Journal*, June 7, 1994.

Peterson, Iver. "Many Bills Are Found Incorrect on Adjustable-Rate Mortgages." *New York Times*, December 12, 1990.

Pickover, Clifford. "What If Every Computer Failed Tomorrow?" *Computer* (September 1991): 152.

Richardson, Lynda. "Software Problem Halts Citibank's Automatic Tellers for 4 Hours." *New York Times*, February 14, 1993.

Rogers, Michael. "Software Makers Battle the Bugs." *Fortune* (February 17, 1986): 83.

Rushing, Angela. "IRS Maps Huge Computer Modernization Plan." *Government Executive* (September 1991): 42–43.

Stix, Gary. "Aging Airways." *Scientific American* (May 1994): 96–104.

Weiser, Mark. "Some Computer Science Issues in Ubiquitous Computing." *Communications of the ACM* 36 (July 1993): 74–84.

Yoder, Stephen K. "When Things Go Wrong." *Wall Street Journal,* November 14, 1994.

Zachary, G. Pascal. "Climbing the Peak: Agony and Ecstasy of 200 Code Writers Beget Windows NT." *Wall Street Journal,* May 26, 1993.

## I. INSIDE RISKS

ACM Committee on Computers and Public Policy. "A Problem-List of Issues Concerning Computers and Public Policy." *Communications of the ACM* 17 (September 1974): 495–503.

Anthes, Gary H. "Phone Outages Traced to Software Updates." *Computerworld* (July 15, 1991): 4.

Arthur, Charles. "Why Computers Could Flip Out in a Tight Corner." *New Scientist* (13 November 1993): 13.

Betts, Mitch. "Glitch Let Scud Beat Patriot." *Computerworld* (May 27, 1991): 109.

———. "Beware of 'Automation Complacency.' " *Computerworld* (May 4, 1992): 87.

Chapman, Gary. "Bugs in the Program." *Communications of the ACM* 33 (March 1990): 251–52.

Cherniavsky, John C. "Software Failures Attract Congressional Attention." *Computing Research News* (January 1990): 4–5.

Cole, Jeff. "Boeing Wants New Jet to Bypass Usual Trials and Fly Ocean Routes." *Wall Street Journal,* July 8, 1994.

Corbató, Fernando J. "On Building Systems That Will Fail." *Communications of the ACM* 34 (September 1991): 72–81.

Corbató, F. J., J. H. Saltzer, and C. T. Clingen. "Multics: The First Seven Years." In *Proceedings of the Spring Joint Computer Conference, May 1972,* 571–583. Montvale, N.J.: AFIPS Press.

De Montalk, J. P. Potocki. "Computer Software in Civil Aircraft." In *Proceedings of the Sixth Annual Conference on Computer Assurance (COMPASS '91),* June 1991, 10–16. Piscataway, N.J.: IEEE Press.

Dorsett, Robert D. "Risks in Aviation." *Communications of the ACM* 37 (January 1994): 154.

———. "Safety in the Air." *Communications of the ACM* 37 (February 1994): 146.

Frenkel, Karen A. "An Interview with Fernando Jose Corbató." *Communications of the ACM* 34 (September 1991): 82–90.

Garfinkel, Simson L. "The Dean of Disaster." *Wired* (December 1993).

Garman, John R. "The 'Bug' Heard 'Round the World." *ACM SIGSOFT Software Engineering Notes* 6 (October 1981): 3–10.

General Accounting Office. *Patriot Missile Defense: Software Problem Led to System Failure at Dhahran, Saudi Arabia.* Washington, D.C.: U.S. General Accounting Office, 1992.

Gruman, Galen. "Major Changes in Federal Software Policy Urged." *IEEE Software* (November 1989): 78–80.

Holusha, John. "Can Boeing's New Baby Fly Financially?" *New York Times,* March 27, 1994.

Homer, Steve. "Battling on with Veteran Computers." *New Scientist* (14 November 1992): 32–35.

Hughes, David. "Raytheon Team Pursues British Contract as U.S. Army Orders Patriot Update." *Aviation Week & Space Technology* (June 3, 1991): 84–85.

———. "Tracking Software Error Likely Reason Patriot Battery Failed to Engage Scud." *Aviation Week & Space Technology* (June 10, 1991).

Lenorovitz, Jeffrey M. "Confusion over Flight Mode May Have Role in A320 Crash." *Aviation Week & Space Technology* (February 3, 1992): 29–30.

———. "French Government Seeks A320 Changes Following Air Inter Crash Report." *Aviation Week & Space Technology* (March 2, 1992): 30–31.

Littlewood, Bev, and Lorenzo Strigini. "The Risks of Software." *Scientific American* (November 1992): 62–75.

Marshall, Eliot. "Fatal Error: How Patriot Overlooked a Scud." *Science* 255 (13 March 1992): 1347.

Matthews, Robert. "Airbus Safety Claim 'Cannot Be Proved.' " *New Scientist* (7 September 1991): 30.

Moxon, Julian. "Will Accidents Always Happen?" *New Scientist* (17 October 1992): 22–23.

Neumann, Peter G. "Inside Risks." *Communications of the ACM,* monthly column.

———. "The N Best (or Worst) Computer-Related Risk Cases." In *Proceedings of the Conference on Computer Assurance (COMPASS '87),* June 1987, xi–xiii. Piscataway, N.J.: IEEE Press.

―――. "The Computer-Related Risk of the Year: Computer Abuse." In *Proceedings of the Conference on Computer Assurance (COMPASS '88)*, June 1988, 8–12. Piscataway, N.J.: IEEE Press.

―――. "The Computer-Related Risk of the Year: Misplaced Trust in Computer Systems." In *Proceedings of the Fourth Annual Conference on Computer Assurance (COMPASS '89)*, 9–13. Piscataway, N.J.: IEEE Press, 1991.

―――. "The Computer-Related Risk of the Year: Distributed Control." In *Proceedings of the Fifth Annual Conference on Computer Assurance (COMPASS '90)*, June 1990, 173–77. Piscataway, N.J.: IEEE Press.

―――. "The Computer-Related Risk of the Year: Weak Links and Correlated Events." In *Proceedings of the Sixth Annual Conference on Computer Assurance (COMPASS '91)*, June 1991, 5–8. Piscataway, N.J.: IEEE Press.

―――. *Computer-Related Risks*. Reading, Mass.: Addison-Wesley, 1994.

O'Lone, Richard G. "Final Assembly of 777 Nears." *Aviation Week & Space Technology* (October 12, 1992): 48–50.

Oster, Clinton V., John S. Strong, and C. Kurt Zorn. *Why Airplanes Crash: Aviation Safety in a Changing World*. New York: Oxford University Press, 1992.

Patel, Tara. "French Fight over Role of Radar in Airline Safety." *New Scientist* (30 January 1993): 20.

Peterson, Ivars. "Phone Glitches and Other Computer Faults." *Science News* 140 (July 6, 1991): 7.

Plant, Norman. "Software in Crisis." *Physics World* (September 1992): 43–47.

Pool, Robert. "Mystery Solved in U.S. Phone Failures." *Nature* 352 (18 July 1991): 178.

Richards, Evelyn. "Writing Software: A Quirky, Labor-Intensive Scramble." *Washington Post*, December 10, 1990.

Schmitt, Eric. "Army Says Computer Failure Linked to Dhahran Deaths Was Like Earlier Error." *New York Times*, June 6, 1991.

Skeel, Robert. "Roundoff Error and the Patriot Missile." *SIAM News* (July 1992): 11.

Skrzycki, Cindy, and John Burgess. "Phone Technology Makes Outages Likely, Experts Say." *Washington Post*, June 28, 1991.

Sparaco, Pierre. "Former Employee Charged in Air Inter A320 Crash." *Aviation Week & Space Technology* (January 25, 1993): 56.

Stix, Gary. "Along for the Ride?" *Scientific American* (July 1991): 94–106.

Stover, Dawn. "The Newest Way to Fly." *Popular Science* (June 1994).

Subcommittee on Investigations and Oversight of the House Committee on Science, Space, and Technology. *Bugs in the Program: Problems in Federal Gov-*

*ernment Computer Software Development and Regulation.* Washington, D.C.: Government Printing Office, 1990.

Waldrop, M. Mitchell. "Congress Finds Bugs in the Software." *Science* (10 November 1989): 753.


## 2. SILENT DEATH

Aeronautics and Space Engineering Board Committee for Review of Oversight Mechanisms for Space Shuttle Flight Software Development Processes. *An Assessment of Space Shuttle Flight Software Processes.* Washington, D.C.: National Academy Press, 1993.

Bhansali, P. V. "Survey of Software Safety Standards Shows Diversity." *Computer* (January 1993): 88–89.

Broad, William J. "New Idea on *Titanic* Sinking Faults Steel as Main Culprit." *New York Times,* September 16, 1993.

Brown, Michael L. "What Is Software Safety and Whose Fault Is It, Anyway." In *Proceedings of the Second Annual Conference on Computer Assurance (COMPASS '87),* June 1987, 70–71. Piscataway, N.J.: IEEE Press.

———. "Software Systems Safety and Human Errors." In *Proceedings of the Conference on Computer Assurance (COMPASS '88),* June 1988, 19–28. Piscataway, N.J.: IEEE Press.

Clayton, Graham. "Europe's Skies Too Busy for Cockpit Computers?" *New Scientist* (13 November 1993): 24.

Covault, Craig. "Mission Control Saved Intelsat Rescue from Software, Checklist Problems." *Aviation Week & Space Technology* (May 25, 1992): 78–79.

Craigen, Dan, Susan Gerhart, and Ted Ralston. "Case Study: Traffic Alert and Collision-Avoidance System." *IEEE Software* (January 1994): 35–37.

Denning, Peter J. "Deadlocks." *American Scientist* 76 (January/February 1988): 11–12.

Fisher, Lawrence M. "Bedside Computers Watch the Vital Signs." *New York Times,* April 26, 1992.

Gannon, Robert. "What Really Sank the Titanic." *Popular Science* (February 1995).

Grossman, Jerome H. "Plugged-in Medicine." *Technology Review* (January 1994): 22–29.

Holden, Constance. "Regulating Software for Medical Devices." *Science* 234 (3 October 1986): 20.

Houston, M. Frank. "What Do the Simple Folk Do? Software Safety in the Cottage Industry." In *Proceedings of the Conference on Computer Assurance (COMPASS '87)*, June 1987, S20–S24. Piscataway, N.J.: IEEE Press.

———. "Directions." In *Proceedings of the Second Annual Conference on Computer Assurance (COMPASS '87)*, June 1987, 121. Piscataway, N.J.: IEEE Press.

Jacky, Jonathan. "Programmed for Disaster: Software Errors That Imperil Lives." *The Sciences* (September/October 1989): 22–27.

———. "Risks in Medical Electronics." *Communications of the ACM* 33 (December 1990): 138.

Joyce, Edward J. " 'Malfunction 54': Unraveling Deadly Medical Mystery of Computerized Accelerator Gone Awry." *American Medical News* (October 3, 1986).

Knight, J. C., and Nancy G. Leveson. "An Experimental Evaluation of the Assumption of Independence in Multiversion Programming." *IEEE Transactions on Software Engineering* 12 (January 1986): 96–109.

Kolkhorst, B. G., and A. J. Macina. "Developing Error-Free Software." In *Proceedings of the Conference on Computer Assurance (COMPASS '88)*, June 1988, 99–107. Piscataway, N.J.: IEEE Press.

Langreth, Robert. "Fail-Safe Skies Coming in the 21st Century?" *Popular Science* (January 1993).

Laprie, Jean-Claude, and Bev Littlewood. "Probabalistic Assessment of Safety-Critical Software: Why and How?" *Communications of the ACM* 35 (February 1992).

Leveson, Nancy G. *Safeware: System Safety in the Computer Age.* Reading, Mass.: Addison-Wesley, 1994.

———. "Software Safety: What, Why, and How." *ACM Computing Surveys* 18 (June 1986): 125–64.

———. "What Is Software Safety?" In *Proceedings of the Conference on Computer Assurance (COMPASS '87)*, June 1987, 74–75. Piscataway, N.J.: IEEE Press.

———. "Software Safety in Embedded Computer Systems." *Communications of the ACM* 34 (February 1991): 34–46.

———. "High-Pressure Steam Engines and Computer Software." *Computer* (October 1994): 65–73.

Leveson, Nancy G., and Clark S. Turner. "An Investigation of the Therac-25 Accidents." *Computer* (July 1993): 18–41.

———. "Therac-25 Revisited." *Computer* (October 1993): 4–5.

Margolis, Nell. "Life-Saving Data Sharing." *Computerworld* (July 26, 1993): 28.

McCrone, John. "Computer Chaos at Medicine's Cutting Edge." *New Scientist* (25 September 1993): 25–29.

Moore-Ede, Martin. "Alert at the Switch." *Technology Review* (October 1993): 52–59.

Neumann, Peter G. "What Is Software Safety?" In *Proceedings of the Conference on Computer Assurance (COMPASS '87)*, June 1987, 76. Piscataway, N.J.: IEEE Press.

Peterson, Ivars. "A Digital Matter of Life and Death." *Science News* 133 (March 12, 1988): 170–71.

Peterson, Margaret. "Advances Cited in Computer-Based Medical Systems Since Therac-25 Accidents." *Computer* (September 1993): 109.

Phillips, Don. "Pilot Blames Plane's System for Problem." *Washington Post*, November 19, 1993.

Richards, Evelyn. "Software's Dangerous Aspect." *Washington Post*, December 9, 1990.

Rifkin, Glenn. "New Momentum for Electronic Patient Records." *New York Times*, May 2, 1993.

Santel, D., C. Trachtmann, and W. Liu. "Formal Safety Analysis and the Software Engineering Process in the Pacemaker Industry." In *Proceedings of the Conference on Computer Assurance (COMPASS '88)*, June 1988, 129–131. Piscataway, N.J.: IEEE Press.

Schneidewind, Norman F., and Ted W. Keller. "Applying Reliability Models to the Space Shuttle." *IEEE Software* (July 1992): 28–33.

Sims, David. "False Alarms Trigger Changes in Collision-Avoidance Code." *IEEE Software* (November 1992): 107–108.

———. "Panel Says NASA Should Strengthen Safety Offices." *IEEE Software* (September 1993): 89–91.

## 3. POWER FAILURE

Archinoff, G. H., R. J. Hohendorf, A. Wassyng, B. Quigley, and M. R. Borsch. "Verification of the Shutdown System Software at the Darlington Nuclear Generating Station." In *Proceedings of the International Conference on Control & Instrumentation in Nuclear Installations*, May 1990.

Arthur, Charles. "Extra Tests Forced on Sizewell's Safety Software." *New Scientist* (6 November 1993): 6–7.

Beardsley, Tim. "SDI: Software Rows." *Nature* 319 (30 January 1986): 345.

Borning, Alan. "Computer System Reliability and Nuclear War." *Communications of the ACM* 30 (February 1987): 112–31.

Chapman, Gary. "Move Over, Nintendo, Here Comes SDI's National Test Bed." *Bulletin of the Atomic Scientists* (November 1988): 12–14.

Charles, Dan. "Unhappy Birthday for Star Wars." *New Scientist* (28 March 1992): 13.

Craigen, Dan, Susan Gerhart, and Ted Ralston. "Case Study: Darlington Nuclear Generating Station." *IEEE Software* (January 1994): 30–32.

Cross, Michael. "Nuclear Power on Britain's Back Burner." *New Scientist* (6 November 1993): 34–39.

Dolnick, Edward. "Can Computers Cope With War?" *Boston Globe*, December 10, 1984.

Everett, Melissa. *Breaking Ranks*. Philadelphia: New Society, 1989.

Gavaghan, Helen. "Professors Say SDI Computers Will Not Work." *New Scientist* (31 October 1985): 14.

Heninger, K., J. Kallander, D. L. Parnas, and J. Shore. *Software Requirements for the A-7E Aircraft*. Washington, D.C.: NRL Report 3876, November 1978.

Humphrey, Watts S. "We Can Program SDI." *IEEE Spectrum* (April 1986): 16.

Lamb, John. "The Bugs in the Star Wars Program." *New Scientist* (21 November 1985): 27–29.

Lin, Herbert. "The Software for Star Wars: An Achilles Heel?" *Technology Review* (July 1985): 16–18.

———. "The Development of Software for Ballistic-Missile Defense." *Scientific American* 253 (December 1985): 46–53.

Marks, Paul. "Faults Highlight Problems of Nuclear Software." *New Scientist* (29 August 1992): 19.

Matras, John R. "Revised Standard Addresses Design Requirements for Computers Used in Nuclear Power-Plant Safety Systems." *Computer* (May 1993): 76–79.

Nelson, Greg, and David Redell. "The Star Wars Computer System." *Abacus* 3 (Winter 1986): 8–20.

Neumann, Peter G. "Managing Complexity in Critical Systems." In *Proceedings of the 1990 ACM Conference on Critical Issues*, November 1990, 47–53. New York: ACM Press.

Palca, Joseph. "SDI Simulations Under Fire." *Nature* 333 (26 May 1988): 286.

Parnas, David L. "A Technique for Software Module Specification with Examples." *Communications of the ACM* 15 (May 1972): 330–36.

———. "On the Criteria to Be Used in Decomposing Systems into Modules." *Communications of the ACM* 15 (December 1972): 1053–58.

———. "Software Aspects of Strategic Defense Systems." *American Scientist* 73 (September/October 1985): 432–40.

————. "SDI: A Violation of Professional Responsibility." *Abacus* 4 (Winter 1987): 46–52.

————. Foreword to *Digital Woes: Why We Should Not Depend on Software*, by Lauren Ruth Wiener, ix–xiii. Reading, Mass.: Addison-Wesley, 1993.

Parnas, David Lorge, and G. J. K. Asmis. "Managing Complexity in Safety-Critical Software." In *Proceedings of the 1990 ACM Conference on Critical Issues*, November 1990, 25–35. New York: ACM Press.

Parnas, D. L., G. J. K. Asmis, and J. Madey "Assessment of Safety-Critical Software in Nuclear Power Plants." *Nuclear Safety* 32 (April-June 1991): 189–98.

Parnas, David L., John van Schouwen, and Shu Po Kwan. "Evaluation of Safety-Critical Software." *Communications of the ACM* 33 (June 1990): 636–48.

Peterson, Ivars. "Finding Fault." *Science News* 139 (February 16, 1991): 104–106.

Rose, Craig D. "SDI Won't Fly, Say Computer Experts." *Electronics* (October 28, 1985): 18–19.

Shore, John. *The Sachertorte Algorithm and Other Antidotes to Computer Anxiety*. New York: Viking, 1985.

————. "Why I Never Met a Programmer I Could Trust." *Communications of the ACM* 31 (April 1988): 372–75.

Waldrop, M. Mitchell. "Resolving the Star Wars Software Dilemma." *Science* 232 (9 May 1986): 710–13. Letters in response: *Science* 233 (25 July 1986): 403–404.

Zraket, C. A. "Uncertainties in Building a Strategic Defense." *Science* 235 (27 March 1987): 1600–1606.

## 4. EXPERIENCE FACTORY

Arnett, Eric H. "Welcome to Hyperwar." *Bulletin of the Atomic Scientists* (September 1992): 14–21.

Arthur, Lowell Jay. "Quick & Dirty." *Computerworld* (December 14, 1992).

Barbacci, Mario H., A. Nico Habermann, and Mary Shaw. "The Software Engineering Institute: Bridging Practice and Potential." *IEEE Software* (November 1985): 4–21.

Basili, Victor, and Scott Green. "Software Process Evolution at the SEL." *IEEE Software* (July 1994): 58–66.

Basili, Victor R., and Harlan D. Mills. "Understanding and Documenting Programs." *IEEE Transactions on Software Engineering* 8 (May 1982): 270–83.

Basili, Victor R., and Robert W. Reiter, Jr. "An Investigation of Human Factors in Software Development." *Computer* (December 1979): 21–38.

———. "A Controlled Experiment Quantitatively Comparing Software Development Approaches." *IEEE Transactions on Software Engineering* 7 (May 1981): 299–312.

Basili, Victor R., Richard W. Selby, and David D. Hutchens. "Experimentation in Software Engineering." *IEEE Transactions on Software Engineering* 12 (July 1986): 733–43.

Basili, Victor R., and Albert J. Turner. "Iterative Enhancement: A Practical Technique for Software Development." *IEEE Transactions on Software Engineering* 1 (December 1975): 390–96.

Basili, Victor R., and Marvin V. Zelkowitz. "Analyzing Medium-Scale Software Development." In *Proceedings of the Third International Conference on Software Engineering,* 1978, 116–123. New York: ACM Press.

Baumert, John. "New SEI Maturity Model Targets Key Practices." *IEEE Software* (November 1991): 78–79.

Curtis, Bill. "Maturity from the User's Point of View." *IEEE Software* (July 1993): 89–90.

Davis, Alan. "Software Lemmingeering." *IEEE Software* (September 1993).

Davis, Joel. *Flyby: The Interplanetary Odyssey of Voyager 2.* New York: Atheneum, 1987.

De Jager, Peter. "Are We Just Plain Lazy." *Computerworld* (February 21, 1994): 85–86.

Endres, Albert. "An Analysis of Errors and Their Causes in System Programs." *IEEE Transactions on Software Engineering,* 1 (June 1975): 140–49.

Glass, Robert L. *Software Conflict: Essays on the Art and Science of Software Engineering.* Englewood Cliffs, N.J.: Prentice-Hall, 1991.

———. "The Software-Research Crisis." *IEEE Software* (November 1994): 42–47.

Gotterbarn, Donald. "SEL 'Experience Factory' Explored at Software Engineering Workshop." *Computer* (February 1993): 117–18.

Grady, Robert B. "Practical Results from Measuring Software Quality." *Communications of the ACM* 36 (November 1993): 62–67.

Gunther, Judith, Suzanne Kantra, and Robert Langreth. "The Digital Warrior." *Popular Science* (September 1994).

Harel, David. "Biting the Silver Bullet: Toward a Brighter Future for System Development." *Computer* (January 1992): 8–19.

Harris, John F. "In Electronic Battlefield Training Exercise, 'Fog of War' Remains Thick." *Washington Post,* April 24, 1994.

Jones, Capers. "The Sociology of Software Measurement." *Computerworld* (August 12, 1991): 61–62.

———. "Sick Software." *Computerworld* (December 13, 1993): 115–16.

Knight, John C., and E. Ann Myers. "An Improved Inspection Technique." *Communications of the ACM* 36 (November 1993): 51–54.

Kocher, Bryan. "Space for Computing." *Communications of the ACM* 32 (October 1989): 1159–60.

Lerner, Eric J. "Classifying the Bugs." *IBM Research Magazine* (Winter 1992): 14–17.

Lieblein, Edward. "The Department of Defense Software Initiative—A Status Report." *Communications of the ACM* 29 (August 1986): 734–44.

Ligezinski, Peter. "Looking for the New Challenge." *Computer* (February 1992): 120.

Linger, Richard C., Harlan D. Mills, and Bernard I. Witt. *Structured Programming: Theory and Practice*. Reading, Mass.: Addison-Wesley, 1979.

Littman, Mark. *Planets Beyond: Discovering the Outer Solar System*. New York: Wiley, 1990.

McSharry, Maureen. "At Workshop, NASA Promotes SEL Process." *IEEE Software* (May 1994): 105–106.

Mills, Harlan D. "Software Development." *IEEE Transactions on Software Engineering* 2 (December 1976): 265–73.

———. "Engineering Discipline for Software Procurement." In *Proceedings of the Conference on Computer Assurance (COMPASS '87)*, June 1987, 1–5. Piscataway, N.J.: IEEE Press.

———. *Software Productivity*. New York: Dorset House, 1988.

Mills, Harlan D., Michael Dyer, and Richard C. Linger. "Cleanroom Software Engineering." *IEEE Software* (September 1987): 19–24.

Myers, Ware. "Debating the Many Ways to Achieve Quality." *IEEE Software* (March 1993): 102–103.

Peterson, Ivars. "Superweapon Software Woes." *Science News* 123 (May 14, 1983): 312–13.

Pierce, Keith R. "Rethinking Academia's Conventional Wisdom." *IEEE Software* (March 1993).

Richards, Evelyn. "Pentagon Finds High-Tech Projects Hard to Manage." *Washington Post*, December 11, 1990.

Saiedian, Hossein, and Richard Kuzara. "SEI Capability Maturity Model's Impact on Contractors." *Computer* (January 1995): 16–26.

Thomas, Martyn, and Frank McGarry. "Top-Down vs. Bottom-Up Process Improvement." *IEEE Software* (July 1994): 12–13.

Topper, Andrew, and Paul Jorgensen. "More Than One Way to Measure Process Maturity." *IEEE Software* (November 1991): 9–10.

Walker, Bruce. "Higher Standards for Software Experiments." *IEEE Software* (September 1994): 6.

Washburn, Mark. *Distant Encounters: The Exploration of Jupiter and Saturn.* New York: Harcourt Brace Jovanovich, 1983.

Wirth, Niklaus. "Program Development by Stepwise Refinement." *Communications of the ACM* 14 (April 1971): 221–27.

———. "A Plea for Lean Software." *Computer* (February 1995): 64–68.

Zelkowitz, Marvin V., and Fletcher J. Buckley. "Are Software Engineering Process Standards Really Necessary?" *Computer* (November 1992): 82–84.

## 5. TIME BOMB

Aiken, Peter, Alice Muntz, and Russ Richards. "DOD Legacy Systems: Reverse Engineering Data Requirements." *Communications of the ACM* 37 (May 1994): 26–41.

Betts, Mitch. "Desktops Veer Toward Year 2000 Crisis." *Computerworld* (July 11, 1994).

Boasson, Maarten. "Exploding Complexity May Be Our Own Fault." *IEEE Software* (March 1993): 12.

Boehm, Barry W. "Software Engineering." *IEEE Transactions on Computers* 25 (December 1976): 1226–41.

Bozman, Jean S. "Tandem Bug's World Tour Thwarted." *Computerworld* (September 9, 1991): 4.

———. "Data Center Maintenance: There's No Time for Downtime." *Computerworld* (April 13, 1992).

Britcher, Robert N., and James J. Craig. "Using Modern Design Practices to Upgrade Aging Software Systems." *IEEE Software* (May 1986): 16–24.

Casey, William. "Fin de Siecle Fear and Loathing." *Washington Post*, December 27, 1993.

———. "Time for the Old Date and Switch." *Washington Post*, January 3, 1994.

Chikofsky, Elliot J. "Application of an Information Systems Analysis and Development Tool to Software Maintenance." In *System Description Methodologies: Proceedings of the IFIP TC2 Conference on System Description Methodologies*, May 1983, 503–14. Amsterdam: North-Holland, 1985.

————. "Software Technology People." *IEEE Software* (March 1988): 8–10.

————. "T Minus 10 and Counting." *IEEE Software* (November 1989): 8.

Chikofsky, Elliot J., and James H. Cross II. "Reverse Engineering and Design Recovery: A Taxonomy." *IEEE Software* (January 1990): 13–17.

Chikofsky, Elliot J., and Wilma M. Osborne. "Fitting Pieces to the Maintenance Puzzle." *IEEE Software* (January 1990): 11–12.

Chikofsky, Elliot J., and Burt L. Rubenstein. "CASE: Reliability Engineering for Information Systems." *IEEE Software* (March 1988): 11–16.

Clarke, Arthur C. *The Ghost from the Grand Banks.* New York: Bantam, 1990.

Cross, James H. "Speaker Cites Standard Data Sets as a Major Challenge Facing Software Reverse Engineering Researchers." *Computer* (November 1993): 83–84.

De Jager, Peter. "Doomsday." *Computerworld* (September 6, 1993).

Denning, Peter J. "About Time." *American Scientist* 78 (July/August 1990): 303–306.

"Disaster Is Closer Than You Think." *IEEE Software* (March 1990): 4.

Doscher, Frank. "Re-Engineering Legacy Systems from the Bottom Up." *CASE Trends* (June 1993): 24–28.

Edelstein, D. Vera, and Salvatore Mamone. "A Standard for Software Maintenance." *Computer* (June 1992): 82–83.

Eliot, Lance B. "Legacy Systems, Legacy Options." *Computerworld* (July 11, 1994).

Farrell, James P. "Date Computations into the Third Millennium." In *Proceedings of the Conference on Computer Assurance (COMPASS '88),* June 1988, 13–18. Piscataway, N.J.: IEEE Press.

Fleming, Richard. "No Bones About It: We Need Software Archaeologists." *Computerworld* (May 27, 1991): 23.

Fuggetta, Alfonso. "A Classification of CASE Technology." *Computer* (December 1993): 25–38.

Gabel, David A. "Technology 1994: Software Engineering." *IEEE Spectrum* (January 1994): 38–41.

Glass, Robert L. "Help! My Software Maintenance Is Out of Control!" *Computerworld* (February 12, 1990).

Grochow, Jerrold M. "Tidal Wave Approaching." *Computerworld* (September 20, 1993).

Guerrero, Carlos A. "Write It Down!" *Computerworld* (October 1, 1990).

Gugliotta, Guy. "Computers Have a Date with a Potentially Messy Destiny." *Washington Post,* February 28, 1995.

Hearn, Steve. "Zero! Zippo! Zilch!" *Computerworld* (February 24, 1992): 87–88.

Hitchens, Randall L. "Dating Problems Now? Wait 'Til the Year 2000." *Computerworld* (January 28, 1991): 23.

Johnson, Marylinda. "Problem Statement Language/Problem Statement Analyzer (PSL/PSA)." In *Proceedings of the Symposium on Application and Assessment of Automated Tools for Software Development*, November 1983, 99–104. Piscataway, N.J.: IEEE Press.

Jones, Capers. "Geriatric Care for Legacy Systems." *Computer* (November 1994): 79.

Letovsky, Stanley, and Elliot Soloway. "Delocalized Plans and Program Comprehension." *IEEE Software* (May 1986): 41–49.

Lorin, Harold. "Limits to Distributed Computing." *Computerworld* (October 28, 1991).

Miller, Joel. "Reverse Engineering: Fair or Foul?" *IEEE Spectrum* (April 1993): 64–65.

Müller, Hausi A. "ICSE 15 Expands Technical Program, Joins Three Smaller Conferences." *IEEE Software* (July 1993): 110.

Murray, Jerome T., and Marilyn J. Murray. *Computers in Crisis: How to Avert the Coming Worldwide Computer Systems Collapse.* Princeton, N.J.: Petrocelli Books, 1984.

Nash, Kim S. "Whipping Worn-Out Code into Shape." *Computerworld* (August 17, 1992): 69–70.

Neumann, Peter G. "The Clock Grows at Midnight." *Communications of the ACM* 34 (January 1991): 170.

———. "Leap-Year Problems." *Communications of the ACM* 35 (June 1992): 162.

Ohms, B. G. "Computer Processing of Dates Outside the Twentieth Century." *IBM Systems Journal* 25, (no. 2) (1986): 244–51.

Peterson, Ivars. "Reviving Software Dinosaurs." *Science News* 144 (August 7, 1993): 88–89.

Pratap, Sesha. "Recycling Software." *Computerworld* (August 10, 1992): 60.

Reisman, Sorel. "COBOL: Common Base, Onerous Legacy." *IEEE Software* (November 1993): 112.

Rettig, Marc. "Nobody Reads Documentation." *Communications of the ACM* 34 (July 1991): 19–24.

Rich, Charles, and Richard C. Waters. "Automatic Programming: Myths and Prospects." *Computer* (August 1988): 40–51.

Samuelson, Pamela. "Reverse-Engineering Someone Else's Software: Is It Legal?" *IEEE Software* (January 1990): 90–96.

Savage, J. A. "HP Establishes Life-Support Center for Aging Software." *Computerworld* (May 6, 1991).

Selfridge, P. G., Richard C. Waters, and Elliot J. Chikofsky. "Challenges to the Field of Reverse Engineering." In *Proceedings of the Working Conference on Reverse Engineering*, May 1993, 144–50. New York: ACM Press.

Sharon, David. "A Classification Scheme for Reverse and Re-Engineering Tools." *CASE Trends*, (April 1993): 31–39.

Soloway, Elliot, Jeannine Pinto, Stan Letovsky, David Littman, and Robin Lampert. "Designing Documentation for Delocalized Plans." *Communications of the ACM* 31 (November 1988): 1259–67.

Waters, Richard C., and Elliot Chikofsky. "Reverse Engineering: Progress Along Many Dimensions." *Communications of the ACM* 37 (May 1994): 22–24.

## 6. SORRY, WRONG NUMBER

Andree, Richard V. "Some Foibles of Computer Arithmetic." *Abacus* 3 (Fall 1985): 62–68.

Borrill, Paul, and Robert G. Stewart. "Standards." *IEEE Micro* (August 1984): 3–6.

Bunch, James R. "Three Decades of Numerical Linear Algebra at Berkeley." *SIAM News* (January 1993): 3.

Cipra, Barry. "How Number Theory Got the Best of the Pentium Chip." *Science* 267 (13 January 1995): 175.

Cody, W. J. "Analysis of Proposals for the Floating-Point Standard." *Computer* (March 1981): 63–69.

Cody, W. J., J. T. Coonen, D. M. Gay, K. Hanson, D. Hough, W. Kahan, R. Karpinski, J. Palmer, F. N. Ris, and D. Stevenson. "A Proposed Radix- and Word-Length-Independent Standard for Floating-Point Arithmetic." *IEEE Micro* 4 (August 1984): 86–100.

Cody, W. J., and D. Hough. "A Proposed Standard for Binary Floating-Point Arithmetic." *Computer* 14 (March 1981): 51–62.

Colonna, Jean-François. "The Subjectivity of Computers." *Communications of the ACM* 36 (August 1993): 15–18.

———. "More on the Subjectivity of Computers." *Communications of the ACM* 37 (June 1994): 89–90.

Coonen, Jerome T. "An Implementation Guide to a Proposed Standard for Floating-Point Arithmetic." *Computer* (January 1980): 68–79.

Corcoran, Elizabeth. "How to Drive a Chipmaker Buggy." *Washington Post*, February 9, 1995.

Corless, R. M. "Six, Lies, and Calculators." *American Mathematical Monthly* (April 1993): 344–50.

Gruenberger, Fred. "How to Handle Numbers with Thousands of Digits, and Why One Might Want To." *Scientific American* (April 1984): 19–26.

Kahan, W. "A Survey of Error Analysis." In *Proceedings of the 1971 IFIP Congress (Information Processing 71)*, 1214–39. Amsterdam: North-Holland, 1972.

——. *And Now for Something Completely Different: The Texas Instruments SR-52*. Berkeley: Electronics Research Laboratory, University of California, Berkeley, 1977 (Memorandum No. YCB/ERL M77/23).

——. *Why Do We Need a Floating-Point Arithmetic Standard?* Berkeley: University of California, 1981.

——. "Mathematics Written in Sand—The HP-15C, Intel 8087, etc." In *1983 Statistical Computing Section Proceedings of the American Statistical Association*, August 1983, 12–26, Alexandria, Va.: American Statistical Association.

Kahan, W., and J. Palmer. "On a Proposed Floating-Point Standard." *ACM SIGNUM Newsletter* (October 1979): 13–21.

Kahan, W., and B. N. Parlett. *Can You Count on Your Calculator?* Berkeley: Electronic Research Laboratory, University of California, 1977 (Memorandum No. UCB/ERL M77/21).

Lakhtakia, Akhlesh, and Richard S. Andrulis, Jr. "Unsuspected Dangers of Extrapolating from Truncated Analyses." *Optics & Photonics News* 2 (June 1991): 8–13.

Moler, Cleve. "A Tale of Two Numbers." *SIAM News* 28 (January 1995.)

Nievergelt, Y. "Numerical Linear Algebra on the HP-28 or How to Lie with Supercalculators." *American Mathematical Monthly* 98 (June/July 1991): 539–43.

Peterson, Ivars. "Can You Count on Your Computer?" *Science News* 122 (July 31, 1982): 72–75.

Shukla, Shyam S., and James F. Rusling. "Analyzing Chemical Data with Computers: Errors and Pitfalls." *Analytical Chemistry* 56 (October 1984): 1347A–68A.

Stein, P. G., and N. D. Winarsky. "How Well Does Your Calculator Calculate?" *RCA Engineer* 29 (January/February 1984): 52–56.

Stevenson, David. "A Proposed Standard for Binary Floating-Point Arithmetic." *Computer* (March 1981): 51–63.

Turner, Peter R. "A History of the Lords of Number-Crunching." *American Mathematical Monthly* (December 1992): 907–16.

Wallich, Paul. "Garbage In, Garbage Out: Simple Geometry Brings Supercomputers to Their Knees." *Scientific American* (December 1990): 126.

## 7. ABSOLUTE PROOF

Ardis, Mark, Victor Basili, Susan Gerhart, Donald Good, David Gries, Richard Kemmerer, Nancy Leveson, David Musser, Peter Neumann, and Friedrich von Henke. "Editorial Process Verification." *Communications of the ACM* 32 (March 1989): 287–90 (including responses by James H. Fetzer and Peter J. Denning).

Barwise, Jon. "Mathematical Proofs of Computer System Correctness." *Notices of the American Mathematical Society* 36 (September 1989): 844–51.

Blum, Manuel. *Designing Programs to Check Their Work*. Report TR-88-009. Berkeley: International Computer Science Institute, 1988.

Blum, Manuel, Alfredo de Santis, Silvio Micali, and Giuseppe Persiano. "Noninteractive Zero-Knowledge Proofs." *SIAM Journal on Computation* 20 (December 1991): 1084–1118.

Blum, Manuel, Michael Luby, and Ronitt Rubinfeld. *Self-Testing/Correcting with Applications to Numerical Problems*. rev. ed. Report TR-91-062. Berkeley: International Computer Science Institute, 1991.

Bowen, Jonathan, and Michael G. Hinchey. "Formal Methods and Safety-Critical Standards." *Computer* (August 1994): 68–71.

Brock, Bishop, and Warren A. Hunt, Jr. "Report on the Formal Specification and Partial Verification of the VIPER Microprocessor." In *Proceedings of the Sixth Annual Conference on Computer Assurance (COMPASS '91)*, June 1991, 91–98. Piscataway, N.J.: IEEE Press.

Buckley, T. F., and P. H. Jesty. "Programming a VIPER." In *Proceedings of the Fourth Annual Conference on Computer Assurance (COMPASS '89)* (June 1989): 84–92. New York: IEEE, 1989.

Clark, Don. "Some Scientists are Angry over Flaw in Pentium Chip, and Intel's Response." *Wall Street Journal*, November 25, 1994.

———. "Intel Finds Pumped-Up Image Offers a Juicy Target in Pentium Brouhaha." *Wall Street Journal*, December 5, 1994.

Colburn, Timothy R. "Program Verification, Defeasible Reasoning, and Two Views of Computer Science." *Minds and Machines* 1 (1991): 97–116.

———. "Computer Science and Philosophy." In *Program Verification: Fundamental Issues in Computer Science*, edited by Timothy R. Colburn, J. H. Fetzer, and Terry L. Rankin, 3–31. Dordrecht, the Netherlands: Kluwer Academic Publishers, 1993.

Corcoran, Elizabeth. "A Flaw Chips Away at Intel's Shiny Image." *Washington Post*, December 2, 1994.

Cullyer, W. J. "Implementing High Integrity Systems: The VIPER Microprocessor." In *Proceedings of the Conference on Computer Assurance (COMPASS '88)*, June 1988, 56–66. Piscataway, N.J.: IEEE Press.

Cullyer, W. J., and W. Wong. "A Formal Approach to Railway Signalling." In *Proceedings of the Fourth Annual Conference on Computer Assurance (COMPASS '90)*, June 1990, 102–108. Piscataway, N.J.: IEEE Press.

de Bruijn, N. G. "Checking Mathematics with Computer Assistance." *Notices of the American Mathematical Society* 38 (January 1991): 8–16.

DeMillo, Richard, R. Lipton, and A. Perlis. "Social Processes and Proofs of Theorems and Programs." *Communications of the ACM* 22 (May 1979): 271–80.

Denning, Peter J. "Beyond Formalism." *American Scientist* 79 (January/February 1991): 8–10.

Dobson, B., and J. E. Randell. "Program Verification: Public Image and Private Reality." *Communications of the ACM* 32 (April 1989): 420–22.

———. "Program Verification." *Communications of the ACM* 32 (September 1989).

Dudley, Richard. "Program Verification." *Notices of the American Mathematical Society* 37 (February 1990): 123–24.

Fetzer, James H. "Program Verification: The Very Idea." *Communications of the ACM* 31 (September 1988): 1048–63.

———. "The Very Idea" (Letters by various authors and responses by Fetzer). *Communications of the ACM* 32 (March 1989): 374–81.

———. "More on the Very Idea" (Letters by various authors and responses by Fetzer). *Communications of the ACM* 32 (April 1989): 506–12.

———. "Another Point of View." *Communications of the ACM* 32 (August 1989): 920–921.

———. "Mathematical Proofs of Computer System Correctness: A Response." *Notices of the American Mathematical Society* 36 (December 1989): 1352–53.

———. "The Final Word on Program Verification." *Notices of the American Mathematical Society* 37 (May/June 1990): 562–63.

———. "Philosophical Aspects of Program Verification." *Minds and Machines* 1 (1991): 197–216.

———. "Program Verification." In *Encyclopedia of Computer Science and Technology*, 237–54. New York: Marcel Dekker, 1993.

Geppert, Linda. "Biology 101 on the Internet: Dissecting the Pentium Bug." *IEEE Spectrum* (February 1995): 16–17.

Gerhart, Susan, Dan Craigen, and Ted Ralston. "Experience with Formal Methods in Critical Systems." *IEEE Software* (January 1994): 21–28.

Glanz, James. "Mathematical Logic Flushes Out the Bugs in Chip Designs." *Science* 267 (20 January 1995): 332–333.

Good, Donald I. "Predicting Computer Behavior." In *Proceedings of the Conference on Computer Assurance (COMPASS '88)*, June 1988, 75–83. Piscataway, N.J.: IEEE Press.

Gries, David. "Why Use Logic? Why Prove Programs Correct?" In *The Science of Programming*, edited by David Gries, 1–5. New York: Springer-Verlag, 1981.

Gruenfeld, Lee. "No Software Guarantees." *Computerworld* (September 30, 1991): 65–66.

Hamilton, Margaret H. "Zero-Defect Software: The Elusive Goal." *IEEE Spectrum* (March 1986): 48–53.

Hill, Richard, Paul T. Conte, Thomas W. Parsons, James Geller, and David A. Nelson. "More on Verification" (Letters). *Communications of the ACM* 32 (July 1989): 790–92.

Hoare, C. A. R. "An Axiomatic Basis for Computer Programming." *Communications of the ACM* 12 (1969).

———. "Programming: Sorcery or Science?" *IEEE Software* (April 1984): 5–16.

———. "Maths Add Safety to Computer Programs." *New Scientist* (18 September 1986): 53–56.

Hodges, Andrew. "Alan Turing: Mathematician and Computer Builder." *New Scientist* (15 September 1983): 789–92.

Kieburtz, Richard. "Formal Software Design Methods Next Step in Improving Quality." *Computing Research News* (January 1991): 14.

Landau, Susan. "Zero Knowledge and the Department of Defense." *Notices of the American Mathematical Society* 35 (January 1988): 5–12.

MacKenzie, Donald. "The Fangs of the VIPER." *Nature* 352 (8 August 1991): 467–68.

Markoff, John. "Flaw Undermines Accuracy of Pentium Chips." *New York Times*, November 24, 1994.

Matthews, Robert. "The Chip with a Sting in Its Tale." *New Scientist* (13 July 1991): 20–21.

McIver, Annabelle. "A Question of Identity." *Nature* 368 (14 April 1994): 589–90.

Mimno, Peter. "Bug-Free Systems: A New Technology for Mathematically Provable Software." *Computerworld* (October 11, 1982): 1–24.

Moore, J. Strother, ed. "Special Issue on System Verification." *Journal of Automated Reasoning* 5 (December 1989): 409–530.

Nelson, David A. "Deductive Program Verification (A Practitioner's Commentary)." *Minds and Machines* 2 (1992): 283–307.

Peterson, Ivars. "Keeping Secrets." *Science News* 130 (August 30, 1986): 140–41.
————. "Holographic Proofs." *Science News* 141 (June 6, 1992): 382–83.
Ralston, T. J., and S. L. Gerhart. "Formal Methods: History, Practice, Trends, and Prognosis." *American Programmer* (May 1991).
Shepherd, David, and Greg Wilson. "Making Chips That Work." *New Scientist* (13 May 1989): 61–64.
Tesler, Edward. "You Can Approximate and Still Be Correct." *Computerworld* (October 28, 1991): 25.
Vijayan, Jaikumar. "Intel Miscalculates Pentium User Backlash." *Computerworld* (December 5, 1994).

## 8. HUMAN ERROR

Abrahams, Paul. "The Role of Failure in Software Design." *Communications of the ACM* 29 (December 1986): 1129–30.
Adams, James L. *Flying Buttresses, Entropy, and O-Rings: The World of an Engineer.* Cambridge, Mass.: Harvard University Press, 1991.
Ambrosio, Johanna. "N.J. Licensing Proposal Draws User Opposition." *Computerworld* (August 19, 1991): 4.
Arthur, Charles. "Software Agents to Roam Phone Network." *New Scientist* (30 July 1994): 19.
Barrie, Douglas. "American Court Raises Software Alarm." *New Scientist* (19 May 1988): 39.
Beder, Sharon. "The Fallible Engineer." *New Scientist* (2 November 1991): 38–42.
Beutel, Richard. "Safeguarding Against Computer Malpractice." *High Technology* (January 1987): 61.
Booker, Ellis. "Early Testing a Must for Big Systems." *Computerworld* (June 10, 1991): 27.
Bradsher, Keith. "How A.T.& T. Accident Snowballed." *New York Times*, January 14, 1991.
Brooks, Frederick P. "No Silver Bullet: Essence and Accidents of Software Engineering." *Computer* 20 (April 1987): 10–19.
Brown, Warren, and Cindy Skrzycki. "Phone Service Problems Are Common, GAO Says." *Washington Post*, March 17, 1993.
Buckley, Fletcher J. "Defining Software Engineering as a Profession." *Computer* (August 1993): 76–78.
Butler, Ricky W., and George B. Finelli. "The Infeasibility of Experimental Quantification of Life-Critical Software Reliability." In *Proceedings of the*

*ACM SIGSOFT '91 Conference on Software for Critical Systems,* December 1991, 66–76. New York: ACM Press.

Cameron, E. Jane, and Yow-Jian Lin. "A Real-Time Transition Model for Analyzing Behavioral Compatibility of Telecommunications Services." In *Proceedings of the ACM SIGSOFT '91 Conference on Software for Critical Systems* (December 1991): 101–111.

Casey, Steven M. *Set Phasers on Stun and Other True Tales of Design, Technology, and Human Error.* Santa Barbara, Calif.: Aegean, 1993.

Chung, Fan R. K. "Reliable Software and Communication I: An Overview." *IEEE Journal on Selected Areas in Communications* 12 (January 1994): 23–32.

Collins, W. Robert, Keith W. Miller, Bethany J. Spielman, and Phillip Wherry. "How Good Is Good Enough? An Ethical Analysis of Software Construction and Use." *Communications of the ACM* 37 (January 1994): 81–91.

Corcoran, Elizabeth. "Racing Light: Can Computer Networks Handle the Traffic?" *Scientific American* (December 1992).

Davis, Paul I. "Certification of Safety-Critical Software by Licensed Software Engineers." *Computer* (December 1992): 72–73.

———. "Computer Society Members Weigh in on Software Safety and Licensure Issues." *Computer* (December 1993): 70–71.

Denning, Peter J. "Human Error and the Search for Blame." *Communications of the ACM* 33 (January 1990): 6–7.

———. "Technology or Management?" *Communications of the ACM* 34 (March 1991): 11–12.

———. "An End and a Beginning." *American Scientist* 81 (September/October 1993): 416–18.

Elmer-Dewitt, Philip. "Ghost in the Machine." *Time* (January 29, 1990): 58–59.

Farrell, Chris. "Survival of the Fittest Technologies." *New Scientist* (6 February 1993): 35–39.

Fordahl, Gregory. "System Reliability Focus of Phone Investigations." *IEEE Software* (November 1991): 79–80.

———. "Bill Would Have FCC Set Reliability Standards." *IEEE Software* (January 1992): 86–87.

Frankston, Bob. "Programming No Longer Enough." *Computerworld* (July 27, 1992).

Gemignani, Michael. "Who's Liable When the Computer's Wrong." *Abacus* 2 (Fall 1984): 58–60.

———. "The Regulation of Software." *Abacus* 5 (Fall 1987): 57–59.

Geyelin, Milo. "Faulty Software Means Business for Litigators." *Wall Street Journal,* January 21, 1994.

Gilman, Hank, and William M. Bulkeley. "Can Software Firms Be Held Responsible When a Program Makes a Costly Error?" *Wall Street Journal*, August 4, 1986.

Gleick, James. "The Telephone Transformed—into Almost Everything." *New York Times Magazine*, May 16, 1993.

Griffeth, Nancy D., and Yow-Jian Lin. "Extending Telecommunications Systems: The Feature-Interaction Problem." *IEEE Computer* (August 1993): 14–18.

Hamer, Mick. "Lessons from a Disastrous Past." *New Scientist* (22/29 December 1990): 72–74.

Herring, Susan Davis. *From the Titanic to the Challenger: An Annotated Bibliography on Technological Failures of the Twentieth Century.* New York: Garland, 1989.

Holste, D. J. "Creating a Network That Heals Itself." *AT&T Technology* 5, no. 2 (1993): 42–47.

Horstmann, Cay. "Arguing Against Certification." *Communications of the ACM* 34 (October 1991): 13.

Horwitt, Elisabeth. "N.Y. Sites Unfazed by Outage." *Computerworld* (September 23, 1991).

———. "No Safety Net for Human Error." *Computerworld* (September 23, 1991): 4.

King, Julia. "When Systems Fail: It's C.Y.A. Time." *Computerworld* (March 30, 1992).

Knight, J. C., and N. G. Leveson. "An Experimental Evaluation of the Assumptions of Independence in Multi-Version Programming." *IEEE Transactions on Software Engineering* 12 (January 1986): 96–109.

Knight, John, and Bev Littlewood. "Critical Task of Writing Dependable Software." *Communications of the ACM* 37 (January 1994): 16–20.

Kornel, Amiel. "User vs. Vendor: Are the Scales Tipping?" *Computerworld* (June 4, 1990).

Laplante, Alice. "Lousy Design Made Easy." *Computerworld* (July 4, 1994): 91–92.

Littlewood, Bev, and Lorenzo Strigini. "Validation of Ultrahigh Dependability for Software-Based Systems." *Communications of the ACM* 36 (November 1993): 69–80.

Lowenstein, Frank. "Software Liability." *Technology Review* (January 1987): 9–10.

Matey, James R. "Good Housekeeping Part I: Remaining Sane in the Face of Inevitable Disasters." *Computers in Physics* 8 (November/December 1994): 642–647.

McCarroll, Thomas. "Failing to Connect." *Time* (September 30, 1991): 51.

Murphy, James R. "Computational Malpractice: Who Is at Fault?" *SIAM News* (September 1988).

Musa, John D. "Software Reliability Measurement." *Journal of Systems and Software* 1 (1980): 223–41.

———. "Software Engineering: The Future of a Profession." *IEEE Software* (February 1985): 55–62.

———. "Tools for Measuring Software Reliability." *IEEE Spectrum* (February 1989): 39–42.

———. "A Software Reliability Engineering Practice." *Computer* (March 1993): 77–79.

Neumann, Peter G. "Certifying Professionals." *Communications of the ACM* 34 (February 1991): 130.

———. "Myths of Dependable Computing: Shooting the Straw Herrings in Midstream." In *Proceedings of the Eighth Annual Conference on Computer Assurance (COMPASS '93)*, June 1993, 1–4. Piscataway, N.J.: IEEE Press.

Nissenbaum, Helen. "Computing and Accountability." *Communications of the ACM* 37 (January 1994): 72–80.

Norman, Donald A. "Commentary: Human Error and the Design of Computer Systems." *Communications of the ACM* 33 (January 1990).

———. *Turn Signals Are the Facial Expressions of Automobiles*. Reading, Mass.: Addison-Wesley, 1992.

———. "Toward Human-Centered Design." *Technology Review* (July 1993): 47–55.

Oz, Effy. "When Professional Standards Are Lax: The CONFIRM Failure and Its Lessons." *Communications of the ACM* 37 (October 1994): 29–36.

Palenski, Ronald, and Bruce Bierhans. "Faulty Software: Problem or Puffery?" *Computerworld* (March 28, 1994): 86–88.

Patterson, William R. "This Isn't It, But Maybe We Do Need a Law." *Computerworld* (September 9, 1991): 23.

Peterson, I. "Software Failure: Counting Up the Risks." *Science News* 140 (December 14, 1991): 388–89.

Petroski, Henry. *To Engineer Is Human: The Role of Failure in Successful Design*. New York: St. Martin's Press, 1985.

———. "Successful Design as Failure Analysis." In *Proceedings of the Conference on Computer Assurance (COMPASS '87)*, June 1987, 46–48. Piscataway, N.J.: IEEE Press.

———. *The Evolution of Useful Things: How Everyday Artifacts—from Forks and Pins to Paper Clips and Zippers—Came to Be as They Are*. New York: Knopf, 1992.

———. "Making Sure." *American Scientist* 80 (March/April 1992): 121–24.

————. "History and Failure." *American Scientist* 80 (November/December 1992): 523–26.

————. "How Designs Evolve." *Technology Review* (January 1993): 50–57.

————. "Predicting Disaster." *American Scientist* 81 (March/April 1993): 110–13.

————. "Failed Promises." *American Scientist* 82 (January/February 1994): 6–9.

————. *Design Paradigms: Case Histories of Error and Judgment in Engineering.* Cambridge, England: Cambridge University Press, 1994.

Piattelli-Palmarini, Massimo. *Inevitable Illusions: How Mistakes of Reason Rule Our Minds.* New York: Wiley, 1994.

Richards, Evelyn. "Society's Demands Push Software to Upper Limits." *Washington Post*, December 9, 1990.

Rogers, Michael. "Can We Trust Our Software?" *Newsweek* (January 29, 1990): 70–73.

Roush, Wayne. "Learning from Technological Disasters." *Technology Review* (August/September 1993): 50–57.

Sagdeev, Roald Z. *The Making of a Soviet Scientist: My Adventures in Nuclear Fusion and Space from Stalin to Star Wars.* New York: Wiley, 1994.

Samuelson, Pamela. "Liability for Defective Electronic Information." *Communications of the ACM* 36 (January 1993): 21–26.

Sterling, Bruce. *The Hacker Crackdown: Law and Disorder on the Electronic Frontier.* New York: Bantam, 1992.

Thimbleby, Harold. "Can Anyone Work the Video?" *New Scientist* (23 February 1991): 48–51.

Vincenti, Walter G. *What Engineers Know and How They Know It: Analytical Studies from Aeronautical History.* Baltimore: Johns Hopkins University Press, 1990.

Waldrop, M. Mitchell. "A Bad Week for Soviet Space Flight." *Science* 241 (16 September 1988): 1429.

————. *"Phobos* at Mars: A Dramatic View—and Then Failure." *Science* 245 (8 September 1989): 1044–45.

Wiener, Lauren. "A Trip Report on SIGSOFT '91." *ACM SIGSOFT Software Engineering Notes* 17 (April 1992): 23–38.

Wilson, Norman L., Jr. "Who Should Pay for a Program Error?" *Computerworld* (September 17, 1984).

Wolkomir, Richard. "A Chronicler of Our Thingamabobs and Doohickeys." *Smithsonian* (October 1993): 133–42.

Zave, Pamela. "Feature Interactions and Formal Specifications in Telecommunications." *Computer* (August 1993): 20–29.

# INDEX

Aiken, Howard, 11
aircraft
  A-7E avionics system, 60–61
  Airbus A320 crashes, 3–5, 7,
    16–18, 20–21
  air traffic control system (FAA),
    103, 119
  automated flight control, 4, 5–7,
    17–23, 49–50, 60–61, 81–82, 193,
    199
  Boeing 777, 21–23
  collision-avoidance systems, 49–50
  F-18 jet fighter, 81–82
  F-16 jet fighter software, 127
  numerical computation errors in
    design, 159–61
  Patriot missiles, 23–24, 155–56
  spacecraft. *See* spacecraft
*Apollo 1*, 50
*Apollo 13*, 205
Apple Computers, 168
Archinoff, Glenn, 71–72, 75–77
Asmis, G.J.K., 72–73, 75
Association for Computing
  Machinery (ACM), 9–10, 12
Atomic Energy of Canada, Limited
  (AECL)
  nuclear power plant shutdown
    systems, 57, 69–77
  Therac-25 radiation therapy
    machine malfunctions, 29,
    31–33, 37–44, 46, 57
AT&T, 136, 207–10, 217, 219, 220
  failure of switching system,
    210–16

ITN and TERESA test and
  monitoring software, 214–15,
  220
Australian National Railways
  Commission, 177
automatic programming, 132–35
automobiles, 171
*Aviation Week & Space Technology*, 154

Basili, Victor R., 90–98, 100–101,
  103–8, 111–12, 121, 181–82
Bell Telephone Laboratories, 11–12,
  216, 219
Blum, Manuel, 187–93
Boeing Company, 21–23
Borger, Frank, 41
Borning, Alan, 65
bridge failures, 202–5
British VIPER chip project, 172,
  174–79, 180, 183
Brooks, Frederick P., Jr., 223–24
Butler, Ricky W., 199–201

*Challenger* explosion, 50, 52, 206
Charter Technologies Limited, 178
Chikofsky, Elliot J., 114–16, 117,
  120–32, 134–37, 140, 141
Chung, Fan, 220
Clarke, Arthur C., 123, 140
Cleanroom software engineering
  process, 98–105
Cohen, Danny, 67
Cohn, Avra, 177–78, 183